USING WRITING
to Teach Mathematics

Andrew Sterrett, editor

MAA Notes and Reports Series

The MAA Notes and Reports Series, started in 1982, addresses a broad range of topics and themes of interest to all who are involved with undergraduate mathematics. The volumes in this series are readable, informative, and useful, and help the mathematical community keep up with developments of importance to mathematics.

MAA Notes

1. Problem Solving in the Mathematics Curriculum, *Alan H. Schoenfeld*, Editor.

2. Recommendations on the Mathematical Preparation of Teachers, *Committee on the Undergraduate Program in Mathematics, Panel on Teacher Training.*

3. Undergraduate Mathematics Education in the People's Republic of China, *Lynn A. Steen*, Editor.

5. American Perspectives on the Fifth International Congress on Mathematical Education, *Warren Page*, Editor.

6. Toward a Lean and Lively Calculus, *Ronald G. Douglas*, Editor.

8. Calculus for a New Century, *Lynn A. Steen*, Editor.

9. Computers and Mathematics: The Use of Computers in Undergraduate Instruction, *D. A. Smith, G. J. Porter, L. C. Leinbach, and R. H. Wenger*, Editors.

10. Guidelines for the Continuing Mathematical Education of Teachers, *Committee on the Mathematical Education of Teachers.*

11. Keys to Improved Instruction by Teaching Assistants and Part-Time Instructors, *Bettye Anne Case*, Editor.

13. Reshaping College Mathematics, *Lynn A. Steen*, Editor.

14. Mathematical Writing, by *Donald E. Knuth, Tracy Larrabee, and Paul M. Roberts.*

15. Discrete Mathematics in the First Two Years, *Anthony Ralston*, Editor.

16. Using Writing to Teach Mathematics, *Andrew Sterrett*, Editor.

17. Priming the Calculus Pump: Innovations and Resources, *Thomas W. Tucker*, Editor.

18. Models for Undergraduate Research in Mathematics, *Lester Senechal*, Editor.

19. Visualization in Teaching and Learning Mathematics, *Steve Cunningham and Walter S. Zimmermann*, Editors.

20. The Laboratory Approach to Teaching Calculus, *L. Carl Leinbach et al.*, Editors.

21. Perspectives on Contemporary Statistics, *David C. Hoaglin and David S. Moore*, Editors.

22. Heeding the Call for Change: Suggestions for Curricular Action, *Lynn A. Steen*, Editor.

23. Statistical Abstract of Undergraduate Programs in the Mathematical Sciences and Computer Science in the United States: 1990–1991 CBMS Survey, *Donald J. Albers, Don O. Loftsgaarden, Donald C. Rung, and Ann E. Watkins.*

24. Symbolic Computation in Undergraduate Mathematics Education, *Zaven A. Karian*, Editor.

25. The Concept of Function: Aspects of Epistemology and Pedagogy, *Guershon Harel and Ed Dubinsky*, Editors.

26. Statistics for the Twenty-First Century, *Florence and Sheldon Gordon*, Editors.

27. Resources for Calculus Collection, Volume 1: Learning by Discovery, *Anita Solow*, Editor.

28. Resources for Calculus Collection, Volume 2: Calculus Problems for a New Century, *Robert Fraga*, Editor.

29. Resources for Calculus Collection, Volume 3: Applications of Calculus, *Philip Straffin*, Editor.

30. Resources for Calculus Collection, Volume 4: Problems for Student Investigations, *Mic Jackson and John R. Ramsay*, Editors.

MAA Reports

These volumes are available from The Mathematical Association of America, 1529 Eighteenth Street, NW, Washington, DC 20036. (202) 387-5200 FAX (202) 265-2384

© 1992 by the Mathematical Association of America

ISBN 0-88385-066-4
Library of Congress Catalog Card Number 90-70793

Printed in the United States of America

Current Printing
10 9 8 7 6 5 4 3 2

CONTENTS

Introduction

Mathematicians Write; Mathematics Students Should, Too
Ann K. Stehney, Center for Communications Research

> This essay argues that, contrary to popular belief, writing is important in mathematics. Faculty members consequently have both a responsibility to help students appreciate the importance of writing in doing mathematics and an opportunity to provide the guidance and practice that will improve students' writing skills and reduce their writing anxiety. The author discusses how mathematics faculty may work with their colleagues to tackle problems of writing at the undergraduate level.

CHAPTER 2

Writing for Educational Objectives in a Calculus Course
Sandra Z. Keith, St. Cloud State University

> This paper discusses the use of writing assignments in a calculus course, frequently the first college-level mathematics course to which a student is exposed. It illustrates how short exploratory assignments provide a powerful way to increase the interactive aspect of learning and, in particular, attempts to show how assignments can be modeled on a catalogue of learning "stages" borrowed from Bloom. The stages, while not necessarily firm, are useful for designing writing assignments as a comprehensive learning sequence. In addition, they can suggest a wide variety of possible assignments that help students go beyond superficial stages of understanding. The ultimate expectation of this approach is that students will begin to create for themselves a value in the learning of mathematics.

CHAPTER 3

Writing in Mathematics: A Plethora of Possibilities
Timothy Sipka, Alma College

> This paper considers a variety of ways that writing assignments can be used to enhance the learning of mathematics. It is not a paper that attempts to introduce the reader to a new and innovative way to teach mathematics; nor is it a paper that attempts to win converts to writing across the curriculum programs. It is simply a paper that acquaints the reader with the utility and versatility of an old and dusty tool—the writing assignment.

Getting Started

A Reply to Questions from Mathematics Colleagues on Writing Across the Curriculum
Emelie Kenney, Wilkes College

> This paper is a response to questions and concerns of mathematics instructors who have heard of and may be interested in using writing in their mathematics courses. An overview of Writing Across the Curriculum principles is given, the application of these principles to the teaching of mathematics is discussed, some ideas for implementation in mathematics courses are provided, and special issues of concern are addressed. Several sample assignments and student responses to some assignments are given.

Writing in Mathematics at Swarthmore: PDCs
Stephen B. Maurer, Swarthmore College

This paper describes the writing component of Swarthmore's Primary Distribution Courses and how that component is implemented in mathematics courses. The program's emphasis is on writing papers and writing them *in the style of the discipline*. The difficulties and benefits of this approach are discussed.

A Writing Program and Its Lesson for Mathematicians
Ann K. Stehney, Center for Communications Research

Since 1983, Wellesley College students have been required to take a writing-intensive course that is offered by faculty from across the college on themes from their own disciplines. This article reports on the issues that were addressed in designing the requirement, the support that is provided for faculty and students in the course, and the benefits derived from the program in its first few years. It includes suggestions for developing such a course based on mathematics and ideas for expository writing assignments that could also be used in existing mathematics courses.

Writing in the Math Classroom; Math in the Writing Class (Or How I Spent My Summer Vacation)
Thomas W. Rishel, Cornell University

Over the past fifteen years or so, I have been using writing assignments in some of my mathematics classes. Further, last year I taught a writing course with a substantial mathematics component. This paper summarizes some of my experiences and draws conclusions from these fifteen years.

A Writing Intensive Mathematics Course
Arthur T. White, Western Michigan University

The course Mathematics 314, Mathematical Proofs, has been designated "writing-intensive" at Western Michigan University. In this essay, the events leading to this designation are recounted, the course is described in some detail, and samples of student writing are provided. The utility of the course, for both "writing to learn" and "learning to write" is discussed.

Writing, Teaching, and Learning in Mathematics: One Set of Experiences
Richard J. Maher, Loyola University of Chicago

This article describes some of the experiences of the Department of Mathematical Sciences at Loyola University of Chicago and of one of its faculty with the relationship between writing, teaching, and learning in mathematics. It begins with a discussion of a seminar course involving the construction and analysis of proofs that has been required of our majors for a number of years; this course involves a good deal of writing. The article then describes the author's experience with the use of writing in courses in statistics and operations research and comments on the use of projects and writing assignments in courses for mathematics majors. After discussing the open question of where writing fits into the overall mathematical sciences curriculum, the article concludes by noting a number of topics that might need to be addressed during any study of the relationship between writing, teaching, and learning in the mathematical sciences.

Technical Writing for Mathematics Projects
J. Douglas Faires and Charles A. Nelson, Youngstown State University

Students writing senior projects in Mathematics and Computer Science need technical writing advice that is not always readily available in their major department. Students in the English department who are pursuing a degree in technical writing need a source of material on which to practice their editorial skills. By designing a program in which the project students use the expertise of the students majoring in technical writing, both needs are served. The project students receive the communication advice and consultation they require and the technical writing students have a source of material that requires their editorial expertise. A version of T$_E$X is used for the formatting procedure, so the students writing the papers are able to easily represent the symbols that are common in technical work and the technical writing students become familiar with this notation. This formatting also gives the final paper a professional appearance that makes a favorable impression on perspective employers and encourages students to give presentations on their projects or to submit the works for publication.

On Grading or "to English or not to English"

But This Is Not An English Class
André Michelle Lubecke, Lander College

This article describes the motivation and rationale behind using writing assignments based on outside readings in an elementary statistics class. It discusses the purpose of the assignments, provides the instructions given to the students along with the outside reading list used, and chronicles the results of completing the assignments from both the instructor's and the students' points of views.

CHAPTER 12

You Can and Should Get Your Students to Write in Sentences
Melvin Henriksen, Harvey Mudd College

A simple program, used by the author for many years to get students to explain their work while writing in sentences is described. Succinctly put, all you have to do is refuse to read work that is not explained and stand fast in the face of strong resistance. The pleasure is worth the pain since before the end of the semester, all of the students are writing in sentences and have a clearer idea of what they are doing.

CHAPTER 13

Three R's for Mathematics Papers—'Riting, Refereeing, and Rewriting
Thomas Q. Sibley, St. John's University

This article discusses a successful incorporation of expository papers in upper division mathematics courses. Papers encourage students to learn how to learn mathematics, to pursue topics of interest to them, to improve their writing skills and to become acquainted with the wealth of mathematical literature. The required student refereeing and rewriting confer educational benefits as well as significantly improving the quality of writing. Through refereeing two papers, each student learns additional topics and reads mathematical prose in a new light. The rewriting enables a deeper, clearer understanding of the topic. Researching, writing and rewriting about mathematics have proved valuable for students of all abilities, amply rewarding the time required of them and the professor.

What do Students Say?

Attempting Mathematics in a Meaningless Language
Martha B. Burton, University of New Hampshire

Beginning students of mathematics need to use the symbol-system of elementary algebra as a language supplementary to their natural, spoken language. Students often find themselves isolated from their own intellectual powers by a presumed necessity of working exclusively in an algebraic language that they find purely formal and empty of meaning. The attempt in this essay is to consider the difficulties students have with the algebraic language, some characteristics of the algebraic language itself, and some possible ways of restoring the connection between natural and algebraic languages so as to allow the student to access and express meaning in mathematical statements.

CHAPTER 15

Using Expressive Writing to Support Mathematics Instruction: Benefits for the Student, Teacher, and Classroom
Barbara Rose, Roberts Wesleyan College

This paper describes my experience using expressive writing activities as an integrated part of a three-credit calculus course at Roberts Wesleyan College, a small liberal arts college in Rochester,

NY. As the students engaged in dialogue journal writing, autobiographical narratives and focused in-class writing, I carefully monitored the experience by collecting and analyzing their writing, their evaluations of the writing experience through both written evaluations and interviews, and my field notes. I found benefits for the student as writer, for the teacher as reader, and for the classroom due to the reader-writer interaction. When students engaged in expressive writing activities, they benefited as they wrote about their feelings about mathematics and the course, the mathematical content covered in the course, their processes of doing mathematics, and their views of the discipline. As I read what students wrote, I became aware of individual needs and common difficulties as well as immediate feedback on the course. The journals also positively influenced the student-teacher interaction and classroom atmosphere; the resulting positive rapport enhanced both personal and academic growth and produced a cooperative and caring climate.

Rewriting Our Stories of Mathematics
Linda Brandau, The University of Calgary

We all have a story to tell of our mathematical lives, a story comprised of our personal school experiences, everyday experiences, and professional experiences. We all have also created a story of mathematics, our own perception of what mathematics is, our own understanding of mathematical concepts. The kind of mathematical life story we have to tell and the kind of story of mathematics we have created, work together to influence what we think of ourselves and the kind of career we choose.

Writing in Mathematics: A Vehicle for Development and Empowerment
Dorothy Buerk, Ithaca College

College freshmen, who would prefer not to take mathematics, enter a Writing Seminar in Mathematics course seeing mathematics as mechanical and rote, as a discipline where their own thinking is inappropriate, and feeling intimidated in its presence. The course is designed to change these students' perceptions of mathematics and of themselves as learners of mathematics. In the process presented the students are first encouraged to reclaim their intuition and common sense in areas requiring quantitative thinking and to reflect on their own mathematical ideas. Then they are encouraged to validate their mathematical ideas using theory, data, and logic. This process of change is described in the context of cognitive-developmental theory. Five mathematical situations with writing components illustrate the process. Students' journals, papers, and written metaphors for mathematics are quoted to demonstrate the empowerment that comes with a changed conception of mathematics.

Two Perspectives on a Writing Intensive Course in Operations Research
Mary Margaret Hart McDonald and Coreen Mett, Radford University

During Fall Semester, 1987, the senior level course in Operations Research was taught as part of the Radford University Writing Intensive Course program. A writing intensive course was defined as one which relies heavily on writing to facilitate learning, while the normal content of the course remains the same. The instructors voluntarily chose to teach upper level courses in this manner. Since courses were not formally designated as writing intensive at registration, students were not self-screened in any way.

As part of the university program in Writing Intensive Courses, each participating instructor spent a week after classes ended in May, 1988, analyzing the experience in a course with this special emphasis on writing. In this case, a student was invited to collaborate in examining the impact of writing in learning Operations Research. The two perspectives were written in parallel, with each of us reviewing the learning process through writing. Mary Margaret discusses with candor the strengths and weaknesses of the journal writing experience. Coreen analyzes the writing incorporated in a semester-long project.

CHAPTER 19

A Writing Fellows Program Meets an Abstract Algebra Class: The Instructor's and the Fellow's Perspective
John O. Kiltinen and Lisa M. Mansfield, Northern Michigan University

In this paper, we discuss the role of writing in the learning of abstract algebra, and tell of the experience at Northern Michigan University with the use of a "Writing Fellows Program" in a junior-level abstract algebra course. The paper consists of three parts. The instructor's perspective is given first, followed by that of the writing fellow, a student with a double major in mathematics and English who had taken the algebra course herself prior to her two semesters of tutorial work with it. As an appendix, we have included a "writing guidelines" document which the instructor prepared for use by the students as part of the project.

On Keeping Journals

Writing Abstracts as a Means of Review
David G. Hartz, College of Wooster

> In an effort to encourage students to regularly review the material, a weekly writing assignment was assigned. The students were required to write an "abstract" of the material covered during the previous week. This assignment forced the students to review and prevented them from lagging behind. The abstracts have been favorably received by both the students and the instructor. The mechanics of this assignment and some of the demands it makes on the instructor are discussed, as well as some of the benefits resulting from it. Examples are given from the students' writing and their reaction to the assignment.

CHAPTER 21

Journals and Essay Examinations in Undergraduate Mathematics
Gary L. Britton, The University of Wisconsin Center-Washington County

> This essay discusses two specific techniques which have been used to get calculus and precalculus students to practice expressing mathematics through writing. The first technique is the use of weekly student journals. The other technique is the use of examination questions which require essay type responses. In order for questions of this type to be beneficial they need to be posed in a special way. Limited success in overcoming the problem of students keeping up with their homework has been achieved by using weekly student journals. The procedures for using the journals, their use in grading, the pedagogical advantages, and student opinions of them are discussed. Too often students don't respond well to essay questions because they don't know what is expected of them, they don't know how much mathematics to include, and they assume the reader is the professor so already knows the material. They are inexperienced at writing in general, and about mathematics in particular. Procedures used in order to overcome these difficulties are presented and sample questions are included.

CHAPTER 22

Weekly Journal Entries—An Effective Tool for Teaching Mathematics
Louis A. Talman, Metropolitan State College

> We, as mathematicians, forget or ignore the power of language, especially written language, as a tool for investigation. This note discusses the results of several semesters' experimentation with weekly journal entries submitted by students in mathematics courses. Use of such journal entries proves to be an effective approach to learning mathematics. The journal entries provide a new and effective channel of communication between student and instructor, as well as an environment where students can't avoid exploring mathematics.

Course Specific, but with Broadly Applicable Ideas

Writing Assignments and Course Content
Joanne E. Snow, Saint Mary's College

> The problem of finding assignments in mathematics classes which are natural and which enhance the learning process can be difficult. Such assignments can serve three purposes: to reinforce the day-to-day class work, to help the student master a particular concept covered in several lectures or examine a few related topics, or to provide an opportunity to learn about material tangential to the course. In this paper, we discuss these purposes and tips about the mechanics of making an assignment and presenting it to the students.

CHAPTER 24

Library and Writing Assignments in an Introductory Calculus Class
John R. Stoughton, Hope College

> The mathematics curriculum acts as a service to many other disciplines—biology, chemistry, physics, engineering, economics—to name only a few. Thus while writing is a very important part of the discipline of mathematics, we require very little of it from our students in the introductory courses. In fact, it is not unusual for a student to write his or her first proof during the junior year.

The transition from the sophomore to the junior year is often difficult for a mathematics major. All too often students have reached a point of no (or at least arduous) return in their college career when they discover that they don't really like mathematics. In this paper, the author discusses his own attempt at introducing writing assignments in the beginning of the mathematics curriculum. As the mathematics community continues its discussions of possible reforms in the calculus curriculum and its teaching, the author would like to add one more idea to the pot.

CHAPTER 25

Teaching Mathematics Within the Writing Curriculum
David T. Burkam, The University of Michigan

The author presents an unusual seminar experience in which mathematical ideas were introduced into a freshmen composition program. Rather than incorporating elements of writing within the traditional mathematics classroom, this course primarily provided a rigorous, introductory writing opportunity against a thematic backdrop of the history and philosophy of mathematics. A brief explanation of the general First Year Seminar Program offered by the Residential College at the University of Michigan is followed by a more thorough presentation of the author's specific contribution to the program, a seminar entitled, "Euclid Alone Has Looked On Beauty Bare": Topics in Mathematical Thought. The main focus of this article remains on the writing assignments (short essays, journal entries, and one longer term paper) rather than an exhaustive description of the readings. By his example, the author hopes to encourage the further development of innovative environments whereby the arbitrary distinctions between the mathematical and the English language can be removed.

CHAPTER 26

Writing About Proof
Keith Hirst, University of Southampton

The major part of this article consists of a commentary on students' writings about proof. In their responses to the assignment they have displayed to a remarkable degree their perceptions of proof, its relations with truth and logic, its authoritarian aspect (manifested in the notion of "proof by intimidation"), relationships with forms of explanation, and the social interactions involved in proving. The written assignment giving rise to their work arose naturally from a class discussion of a particular proof of a conjecture arising from a number pattern, making their accounts of the process personal and immediate.

CHAPTER 27

Using Writing to Improve Student Learning of Statistics
Robert W. Hayden, Plymouth State College

This paper discusses student writing assignments (and my goals for same) in two applied statistics courses, one an ordinary Introduction to Statistics and the other an applied regression course locally named Applied Statistics Using the Computer. What I have done can most readily be extended to other courses in which mathematics is applied to the world around us. It can less readily be extended to courses in pure mathematics or to courses devoted primarily to computational techniques or algebraic manipulations.

Integrating Writing into the History of Mathematics
Dorothy Goldberg, Kean College of New Jersey

The history of mathematics is an ideal mathematics course to use to provide opportunities for students to develop their writing competency.

Following guidelines provided by the Kean College of New Jersey Writing Emphasis Committee, a sequence of writing assignments was developed for this course. Most assignments were graded, but usually students first edited each other's paper.

Depending on whether students took the final examination or wrote a term report, 38.5% or 55% of the grade was for writing.

Writing to Learn and Communicate Mathematics: An Assignment in Abstract Algebra
Anne E. Brown, Saint Mary's College

Two premises about the use of writing assignments in mathematics courses form the foundation for this essay. The first is that when students are asked to organize, summarize, and illustrate in writing the concepts they are studying, the learning of mathematics is enhanced. The second is that it is important to encourage mathematics students to develop their ability to communicate mathematical ideas accurately and effectively to a specified audience.

CHAPTER 30

Writing in a Non-Euclidean Geometry Course
Richard S. Millman, Wright State University

Writing intensive courses are appearing as a integral part of "writing across the curriculum." These courses are not just restricted to the humanities and social sciences though, but also are to be given in the student's major. There is skepticism on the part of mathematics faculty because many do not see what a writing intensive advanced mathematics course might be like. This paper outlines a pilot project in which a non-Euclidean geometry course was taught as a writing intensive one. Each student was asked to write an article on some aspect of geometry either covered in the class or of outside interest. The project was in a multidraft format. The purpose of this article is to describe the project, the student reaction to it, and both the benefits and the difficulties inherent in such an approach.

CHAPTER 31

The Essay as a Cognitive Map
James V. Rauff, Milliken University

Do students form a cognitive map of the content of a mathematics course? If so, what is the nature of that map? This paper presents the results of a writing assignment that required the students in a *Foundations of Mathematics* course to present their understanding of how the topics of the course formed a cohesive whole. It is suggested that comprehensive essays provide a good way of getting a glimpse of how the student organizes mathematical material for understanding.

FOREWORD

This collection of essays on the use of writing to teach mathematics is an outgrowth of sessions of contributed papers presented at the 1988 and 1989 Annual Meetings of the Mathematical Association of America. The MAA allocated two and one-half hours to organize the first of these sessions at the 1988 Atlanta Meeting, time enough for ten papers. So many papers were contributed, however, that the MAA extended the time to nine hours over three sessions. I estimated the attendance at those sessions at well over 400, 150 of whom were present at the beginning of an 8:00 a.m. session. Several papers presented at the Atlanta meeting were published in *Writing to Learn Mathematics and Science,* published in the summer of 1989 by the Teachers College Press, Columbia University. Two recent articles in the MAA's *College Mathematics Journal,* "Learning Mathematics Through Writing: Some Guidelines" by J. J. Price (vol. 20, no. 5, November 1989) and "What's an Assignment Like You Doing in a Class Like This?" by George D. Gopen and David A. Smith (vol. 21, no. 1, January 1990) provide additional experiences with writing in mathematics classes. Two MAA publications also will be of interest, *Writing Mathematics Well* by Leonard Gillman and *Mathematical Writing* (MAA Notes 14) by Donald E. Knuth, Tracy Larrabee, and Paul Roberts.

Because of the widespread interest in writing that was exhibited in Atlanta, the MAA asked Gerald Bryce, mathematics professor at Hampden-Sydney College, to organize a similar session at the Phoenix Annual meeting in 1989. More than 200 persons attended the presentation of the 20 papers given there, and this continuing interest in writing to teach mathematics led the MAA to ask me to edit this volume.

I wanted essays for this volume that contained ideas that are easily transported to other institutions and to other courses, and I am confident that each of the essays selected meets that criterion. The writers describe their experiences with Writing Across the Curriculum, with journals and other forms of expressive writing, and with specific courses. They offer specific advice on getting started with writing programs and on routine matters such as grading, correcting grammar, and the importance of rewriting. Several essays describe student reaction to writing in mathematics classes and how to involve students in reading and grading the work of others.

In the first essay of the Introduction, "Mathematicians Write: Mathematics Students Should Too," Ann Stehney offers an excellent rationale for requiring students to write more frequently in their mathematics courses. Then Sandra Keith discusses a theoretical framework (Bloom's Taxonomy) in which to consider the validity of writing assignments and Timothy Sipka actually does describe a "plethora of possibilities."

For many mathematicians, the first step is the most difficult one in requiring students to write about mathematics. We think that we must have as much knowledge as our colleagues in the English Department in order to help students write well, but those colleagues claim to have had little or no formal training that enables them to correct student papers. They claim to be experts on Milton or Shakespeare rather than on Newton or Hilbert. I am not completely comfortable in presenting this argument, but my own experience has been that I have been able to help most students in spite of my limited knowledge of syntax and grammar.

In the first essay in "Getting Started," Emelie Kenney provides thoughtful responses to questions that prevent mathematicians from assigning significant amounts of written work. Stephen Maurer and Ann Stehney describe imaginative Writing Across the Curriculum programs; Thomas Rishel, Arthur White, and Richard Maher describe their individual experiences in getting started. Douglas Faires and Charles Nelson report on an unusual cooperative venture between graduate students in English and students in the mathematical sciences.

Can mathematicians help students improve their writing skills? André Lubecke, Melvin Henriksen, and Thomas Sibley answer affirmatively and then provide some useful suggestions.

Of course, students have many different reactions to the variety of assignments that require them to write in mathematics courses. In "Attempting Mathematics in a Meaningless Language," Martha Burton describes the difficulties that many students have with *our* language. Barbara Rose, Linda Brandau, and Dorothy Buerk report vivid student reactions to mathematics as well as to writing in a mathematics course. John Kiltinen and Coreen Mett have student coauthors who describe their involvement with writing programs at their institutions.

At a recent meeting of the Ohio Section of the MAA, David Hartz explained his procedure for collecting journals in which students summarize the work of the previous week. He requires that students explain fundamental concepts in their own words rather than list formulas. I used this journal technique with a second semester calculus class, and student reaction convinced me that I should collect similar journals in all my classes. Gary Britton and Louis Talman also provide their perspectives on the value of having students keep journals.

The remaining authors describe the writing assignments that they make in specific courses, but their experiences are readily applied to other courses. For example, John Stoughton describes two assignments that require students to write in a calculus course, but his ideas are readily transported to other mathematics courses. Virtually all the ideas described in the essays that appear to relate to specific courses have applications in other mathematics courses as well.

When I attend a conference on pedagogy, I always consider it a success if I learn one new idea that I can apply in my own teaching. I will be greatly surprised if everyone doesn't find in this volume a "plethora of possibilities" that would enhance students' understanding and appreciation of mathematics.

I wish to express my gratitude to Dr. David Hull, Professor of Mathematics at Ohio Wesleyan University, and Dr. Richard Kraus, Professor of English at Denison University, for their thoughtful appraisals of the many manuscripts received and to Dr. Peter Renz, Siobhán Chamberlin, and Beverly Ruedi at MAA Headquarters for their dedication and skill in putting this attractive volume together in such a short time.

Andrew Sterrett
Emeritus Professor of Mathematical Sciences
Denison University, Granville, Ohio
Interim Associate Director for Programs
Mathematical Association of America, Washington, DC

... The day will come, I believe, when the value of writing to learn will be universally acknowledged.

Reuben Hersh
"A Mathematician's Perspective"
Writing to Learn Mathematics and Science

In all other disciplines, some writing—term papers, book reviews, lab reports, etc.—is required. Is mathematics really so special? I don't think so. We have simply neglected an important part of our students' education. We pay the penalty in frustration when we find our math majors practically illiterate in the language of mathematics. Can you imagine students who have taken eight years of French and still cannot write or speak a simple sentence? If my students parachuted into Algebraland, most of them would starve because they could not speak the language.

J. J. Price
"Learning Mathematics through Writing: Some Guidelines"
College Mathematics Journal, November 1989

INTRODUCTION

Mathematicians Write; Mathematics Students Should Too

Ann K. Stehney
Center for Communications Research, Princeton, New Jersey

At high school career fairs, students want to know what a mathematician does all day. My first answer is, "I write." I do other things, of course, like thinking and reading and listening and discussing and computing and lecturing, but I answer as I do to warn students that writing may be important to their work when they don't expect it. At the Center for Communications Research, the only tangible products of my work are the technical reports that I write to disseminate my ideas. I have also published research papers in pure mathematics, expository articles, conference proceedings, reviews of papers and books, editor's notes for collections of articles, and essays like this one. My professional writing that is not meant for publication includes referee's reports, committee reports, grant proposals, memoranda, letters of recommendation, and other correspondence.

I write now. I wrote as a faculty member. I wrote as an administrator. In fact, the only time in my adult life when I did not write was during graduate school, when I was preparing for a career in which I would be writing.

I probably write more than the average mathematician, but not because I have more to say or because it's easier for me. I am not a prolific writer, and after years of experience, I still find writing a slow process of successive approximation. It's just that, unlike other people, I don't try to avoid it. With instruction and practice, I have become less anxious about writing than most people are. I have conquered an aversion to confronting my own rough drafts and a reluctance to asking friends and colleagues for their criticism. I have confidence that by reading and revising, I can produce a piece that conveys what I mean in a fairly clear way.

It would be a disservice to let students embark on a technical career with the assumption that they can substitute symbols and equations and technical jargon for real sentences in honest English words.

According to the stereotype, mathematicians cannot communicate with ordinary people, nor do they have to. Mathematics is identified with the specialized writing of research journals, with its stilted language and terse, unadorned style, in the minds of our students as well as of the general public. But I would argue that such writing is irrelevant to the purposes of this volume. It would be easy for the practiced writer to imitate, so it is not something that needs to be learned. In addition, most of the reading and writing that most students will do in their professional lives will not involve research journals. Few will become PhD mathematicians; few PhD's will publish any research beyond a thesis; and few research papers are read by any number of people.

If so, how does the field develop? How is new mathematics shared? Essentially in the ways that old mathematics is passed on: through lectures, books, and expository or review articles, rarely from the original written sources. Contrary to popular belief, there is a need for good exposition in all areas of the mathematical sciences. Even the technical reports of my own applied mathematics are much more expository in nature than a journal article would be:

I include background material, motivation, details, insight if possible, and even approaches that did not succeed. The paper is an opportunity to share my understanding as well as to take credit for the work. With these examples in mind, it would be a disservice to let students embark on a technical career with the assumption that they can substitute symbols and equations and technical jargon for real sentences in honest English words.

Early in this century, H. G. Wells praised the "efficient mathematical teaching" of modern education. In *Mankind in the Making* [1], he wrote, "The arithmetic (without Arabic numerals, be it remembered) and the geometry of the mediæval quadrivium were astonishingly clumsy and ineffectual instruments in comparison with the apparatus of modern mathematical method . . . [T]he new mathematics is a sort of supplement to language, affording the means of thought about form and quantity and a means of expression more exact, compact, and ready than ordinary language." But as much as he valued precision and brevity, Wells did not mean that the new mathematics could take the place of ordinary language. In his scheme for initiating a citizen of the modern world, he placed "a thorough study of English as a culture language" and "a sound training in prose composition" ahead of "just as much of mathematics as one can get in." As he saw it, the "pressing business" of education is "to widen the range of intercourse . . . We do not progress far with our thoughts unless we throw them out into objective existence by means of words, diagrams, models, trial essays."

While the rest of the population deals with "math anxiety," many mathematicians have to deal with writing anxiety. This is unfortunate, because writing is important to learning: thoughts are developed and refined in the process of writing. Other contributors to this volume will discuss the connections between writing and thinking, and how writing can aid a student's learning of mathematics. I will be concerned here with the related question of how mathematicians can aid a student's learning of writing. My assumptions are these:

> Mathematicians write. They write to propose, to inform, to ask, to sell, to explain, and to plead, as well as to record.

> Good writing is better than poor writing, for the sake of mathematics. The goal is to develop one's ideas and communicate them simply without gaps, jumps, ambiguities, or distracting errors of standard written English.

> Clear and effective prose does not come naturally to many people, but guidance and practice help. Students' writing skills can be improved; their anxiety about writing can be reduced.

While the rest of the population deals with "math anxiety," many mathematicians have to deal with writing anxiety.

How can faculty members in mathematics work with their colleagues in a coordinated attack on problems of writing at the undergraduate level? I have already argued that we mathematicians are responsible for making students understand that writing is important to us. In addition, a writing program would be enriched by the involvement of instructors from different disciplines. Someone with an interest in mathematics and a sense of what mathematicians do can devise exercises whose form and content are inspired by working in the field. The rest of this article will discuss some suggestions for involvement that could be used alone or in combination.

Include Writing Assignments in Mathematics Courses

Use ideas from this volume to get started. Some time ago, I assigned expository writing in my classes without being properly prepared; the results were frankly disappointing. Students who were fulfilling a math or science requirement could not believe that the standards for the course included standards for writing; mathematics majors resented having to write in a geometry course. A senior in an independent study course wrote so poorly that I did not know how to deal with her paper, and an honors student assumed that my comments on her exposition were intended as a criticism of her thinking. I could not single-handedly mount an assault on writing problems and cover a reasonable syllabus. I hope that the recent attention on writing in the disciplines will give others the guidance and courage and community support they need to follow through with writing assignments in mathematics courses.

I hope that the recent attention on writing in the disciplines will give others the guidance and courage and community support they need to follow through with writing assignments in mathematics courses.

Be Involved in Writing Courses Taught by Others

When a member of the Wellesley College English Department offered a course called "Exposition for Experts" for upperclass students to write in their own fields, she asked the Mathematics Department to compile a list of words that should be used with care in a mathematics paper. We included terms for logical relationships, common words whose nontechnical usage should be avoided in technical papers (category, function, group, limit, series, simple, etc.) and colorful words that have acquired mathematical meanings (bundle, germ, radical, soul, surgery, etc.).

This was an easy way to contribute to someone else's writing course. We can also be role models for the students. We can articulate the ways that thinking and writing interact in our work. We can help devise assignments. We can provide resources, including examples of good, poor, and mediocre writing, in various genres and styles, as well as lists of reference works. While ideas and materials could be widely shared, thus increasing the benefits without much additional effort, working with a colleague on his course could be a time-consuming task that is not appreciated by the administration or recognized as a worthwhile scholarly activity at review time. As with all the suggestions on this list, individuals are more likely to participate if there is an institution-wide commitment to such matters.

Team-Teach a Course with an Experienced Writing Instructor

This is a more extensive version of the suggestion above, with the benefit of official recognition. The mathematician is a full partner in preparing for the course, teaching in the classroom, meeting with students, and reading their work.

This option is expensive if the course is counted toward each instructor's normal teaching load. It should be regarded as a temporary situation, part of the on-going training of both instructors and an investment in the future, when each instructor can carry on alone. The administration might justify the course on this basis, or the team-taught course might qualify for special funding. It might even be a one-time contribution that two dedicated and altruistic faculty would be willing to make, to team-teach two sections of the course and each get "credit" for one.

Another potential obstacle is a resistance to team-teaching altogether. Many faculty members, regarding the classroom as their private domain, do not like any visitors or observers. I would hope that this problem might not arise if each member of the team were there by virtue of his own expertise in one area and admitted inexperience in the other.

Teach a Writing Course

This may sound radical, but I believe it is ultimately the most effective and practical approach for an individual to take. It is a natural sequel to the suggestion above, or it could follow training workshops instead of classroom apprenticeships. It assumes that any faculty member who volunteers to teach such a course writes with a minimum of fear, recognizes good writing and poor writing, can learn to identify weaknesses explicitly, and will work to acquire strategies for helping students improve their writing.

I believe that the success of such a course will depend on the instructor's avoiding a hidden agenda to cover a semester's worth of mathematics. Students will, of course, be free to learn mathematics during the course, as well as learning about mathematicians and the history of mathematics and anything else they read. It is also permissible for the teacher to have a hidden agenda to teach clear thinking through clear writing and to help students hone their ideas as they write. On the other hand, the course should not be meant as an easy way for mathematics majors to satisfy a writing requirement. While good mathematics should be encouraged and acknowledged, students should understand that they cannot get through the course by substituting good mathematics for poor writing.

Support a Writing Requirement

Every student could benefit from guidance and practice in expository writing at the college level. If a composition course is required without exception, it will not be viewed as remedial or punitive. If it can involve a theme of interest to the student, the requirement need not be a hardship.

Every student could benefit from guidance and practice in expository writing at the college level. If a composition course is required without exception, it will not be viewed as remedial or punitive.

A new requirement cannot come about without a willingness to adjust to change. In my experience, arguments among the faculty about degree requirements are motivated by philosophical and political concerns as much as by academic ones. Some people want to give students as much latitude as possible in designing their undergraduate program, and so they oppose any additional structure, especially in the requirements for graduation. Others may feel their department is threatened by a requirement that guarantees enrollments to someone else (or conversely, they might expect a windfall of increased enrollments). Or administrators might balk at offering numerous courses with limited enrollments and expect the faculty to make up for it in other classes. Finally, the burden of offering a serious writing program may be too much for a single department to shoulder, especially if it has a large load of remedial or English as a second language courses.

If the campus community can reach a consensus that a writing requirement makes sense for their institution, the faculty can share the responsibility of offering courses that fulfill the requirement. Short of a requirement, there are things that individuals can do for the sake of their own students' writing—mathematicians included.

Short of a requirement, there are things that individuals can do for the sake of their own students' writing—mathematicians included.

ACKNOWLEDGEMENTS   I wish to thank my colleagues Leonard Charlap and David deGeorge for their comments on earlier drafts of this article and related discussions of controversial topics. I am also grateful to Bryn Mawr College for a grueling year of required Freshman Composition that reduced my writing anxiety to manageable levels.

REFERENCE

1.   H. G. Wells, *Mankind in the Making*, Charles Scribner's Sons, New York, 1904: 191–192 and 201–202.

Writing for Educational Objectives in a Calculus Course

Sandra Z. Keith
St. Cloud State University, St. Cloud, Minnesota

> Alice attended to all these directions, and explained as well as she could, that she had lost her way.
>
> "I don't know what you mean by *your* way," said the Queen: "all the ways about here belong to me—but why did you come out here at all?" she added in a kinder tone. "Curtsey while you're thinking what to say. It saves time."
>
> Alice wondered a little at this, but she was too much in awe of the Queen to disbelieve it. "I'll try it when I go home," she thought to herself, "the next time I'm a little late for dinner."
>
> "It's time for you to answer now," the Queen said looking at her watch: "open your mouth a *little* wider when you speak, and always say, 'your Majesty.'"
>
> *Through the Looking Glass,* Lewis Carroll

Alice might have stumbled into a calculus class. Students in calculus receive very little encouragement to communicate to us "their way" rather than "our way." Time pressure pervades the classroom, and students seem to want to "think later," when they go home. On exams, students are suddenly confronted with the need to "answer now," and they well might feel they must "open their mouths a little wider," and give the teacher what that mysterious patron of the subject "wants," or else stifle a thought about "going home to dinner." In this excerpt, it is also revealing that Alice probably does not spend any time thinking at all, about her way or the Red Queen's way, for that matter, so much as wondering.

In real life, however well-intentioned we have been in these past thirty years, calculus has become a "Wonderland." Nationally, only 40% of our students pass this course, the "gateway to science," with a grade of D or better. On the one hand, students have never seemed so unprepared in background work or so casually ignorant of common study skills. On the other hand, we are beginning to admit that calculus as a course has become overcrowded with information, as new applications creep in and outdated methods resist being weeded out. We face the criticism that calculus contains far too many topics that computer classes and pocket calculators make obsolete. There is no clear-cut solution to the problem, but the exhausted teacher might well wonder how anything can be added to a calculus course. This has been a common response against the use of writing assignments in mathematics.

A serious reconsideration of values is in order in the calculus classroom.

The nagging fact remains, however, that a tight schedule requires a "top-down" teaching style, the style which is so characteristic of a mathematics class, with teacher as preacher and students as the tested population. It has become apparent, sadly, that this style is failing to engage our students these days, and students create their own abuses to the system by making it the responsibility of the instructor to make the subject easy for them. An implicit policy between students and teachers of settling for the students'

understanding some basic, precariously-learned rote skills sometimes creeps on us. When we comply, we reduce calculus to a course in algebraic manipulations, the lowest common denominator of learning, and a falling barometer of student understanding. The result is that students are failing to grasp conceptually what calculus is about; the rote skills are lost one year later. We are all paying the price for this. To keep mathematics alive as a subject in college, we must be prepared to create a more vigorous, interactive classroom environment. Years ago (and in other nations), this environment was (and is) achieved by long hours of homework. These days, however, many high school teachers do not enforce homework and students are lacking in elementary study habits to the extent that they appear to have no feeling for what it means to "know" something. The complaint then, that writing assignments take time and energy, must necessarily force us to rethink once again those curricular restrictions we are under, where departmental controls and tight schedules provide little room for the kind of discursive, responsive approach to teaching that we need. In particular, a serious reconsideration of values is in order in the calculus classroom.

Purpose of Writing Assigments

My way of dealing with this problem, at the large state university at which I teach, is to rely on "exploratory" writing assignments. I define these frequently short, impromptu assignments to be assignments which aim to focus a student's understanding by trial-and-error, in the student's own words, with feedback from the instructor. I favor these exercises partly because they can be inserted into a tight curriculum. More specifically, these exercises serve several important purposes.

Frequently we may think we understand something when we only recognize it; we confuse familiarity with understanding. This becomes obvious when we have to explain it in writing.

1. They allow students to explore and organize the mathematics presented in their own terms; they allow for invention, and they facilitate both learning and retention. They enable students to come to an understanding of what it means to "know" and to develop a method for learning that aims to achieve this state of "knowing" which is generally far more thorough than anything they have imagined. Frequently we may think we understand something when we only recognize it; we confuse familiarity with understanding. This becomes obvious when we have to explain it in writing.

2. They give students experience in writing and understanding writing—as it might be presented by the teacher, and in textbooks. They allow students to read and process the writing of other students, and they allow students to "talk" mathematics with more ease, and to develop a language for asking questions.

3. They provide the instructor with an open window on where the class stands, and immediate feedback on how to teach to the problems of the class or where reteaching is necessary, prior to an exam. Frequently these assignments allow students to be "twice-tested."

4. The in-class evaluation of writing assignments stimulates meaningful discussions of mathematics as a language and strategies for learning it. The assignments break down fear in the classroom, and nuture a more open environment for asking questions.

5. They create substantial hurdles for even the best students, and thus broaden opportunities for conceptual growth for everyone. All students become more consistently engaged and more self-reliant.

6. Most importantly, writing assignments enhance the intrinsic *value* of mathematics for the student. That is, they enable us to create a value in the doing and learning of mathematics—its terminology, language, rigor, elegance, and invention. When a student asks, "Why are we doing this?" my traditional response has been to pull out applied examples as if from a closet, and yet these examples never appear to satisfy the student. Rather, the student seems to be asking, "What is the value to me in this method?" The question, "What good is it?" really means, "Why should I admire or love this subject?" By inviting students to become personally involved in the learning environment, writing assignments help transmit the value of doing mathematics for its own sake. To me, the main value in mathematics is its ability to broaden our sense of learning and knowing and thinking in depth, and writing assignments are consonant with this value.

The Practical Use of Writing Assignments

When using these exercises, I frequently devote a 10–15 minute period of the class to writing-on-the-spot, individually, or in groups. Longer, organizational themes, I assign as overnight exercises. The evaluation of these assignments is crucial in the process of using these assignments. I have requested an overhead projector as a permanent fixture in my classroom, and I often reproduce students' writing on transparencies, or the students write directly on the transparencies. Some of the writing is exhibited, and we make comparisons and contrasts. Concepts and writing play dual roles; concepts are evaluated through writing and writing through concepts. Every effort is made to understand what the student was thinking at the time of writing, so that no shame and embarrassment is involved; generally there is something good to say about almost every piece of writing, and interestingly, students' errors fall into categories, which is enlightening to students who often see their failure to understand as uniquely "dumb." Students appear to understand what I am trying to do, and I have always found student reaction to these assignments extremely positive; my students generally appreciate anything that has a grade attached, however minimal it may be. While I do not generally grade, when I do, I set the terms that the students must communicate with me, so that grading goes fast and easily; I do not labor to understand the student. I regard grammar and spelling as a secondary concern in so far as I can understand what students are saying. (On the correct spelling of mathematical terms I make a stringent exception.) Continually correcting grammar and spelling tends to intimidate students into using a small vocabulary and taking fewer risks. Above all, I enjoy designing assignments, as they keep a course fresh and challenging to me as well as to the students. When I find students staring blankly at me, I often ask them to, in a 10-minute period, define a term, state the theory we have been doing, provide an example, or predict what's coming: The question, "why are we studying this?" is seized upon as an excellent group assignment, and so on.

I enjoy designing assignments, as they keep a course fresh and challenging to me as well as to the students.

A Heursitic for a Writing Assignment

The cutting edge in the use of writing assignments is the problem of how to respond to students' writing in a way which is productive in the long run. When we cannot necessarily give students personal attention, how can we help students to get an organized, overall sense of how to learn, and in a such a way that the entire class benefits? Without follow-through experiments and testing on the subject, how can we be sure students are actually improving with writing assignments? An approach to the problem of seeing the validity of writing assignments occurred to me in a writing-across-the-curriculum workshop, where I was introduced to the educational objectives in Bloom's *Taxonomy* (Bloom is probably familiar to educators.) Bloom is ultimately concerned with creating a value system in the learning process. But at a more accessible level, he provides a sequence of educational objectives which have encouraged me to focus my assignments toward some sequenced view of learning; these objectives are synopsized here. All of these objectives will seem familiar, although the boundaries are somewhat fuzzy.

1. *Knowledge.* Remembering, recalling, the lowest level of learning.

2. *Comprehension.* The ability to grasp the meaning as shown by translating material from one form to another, interpreting material—explaining or summarizing. These learning outcomes are one step beyond simple remembering, but they represent the lowest level of understanding.

3. *Application.* The ability to use the material in new and concrete settings, to use rules, methods, concepts, principles, laws, theories. At this level, learning outcomes go beyond comprehension.

4. *Analysis.* The ability to break down material into its component parts, to anyalze the relationship between the parts, and to recognize organizational principles involved. These learning outcomes represent a higher intellectual level than comprehension and application.

5. *Synthesis.* The ability to put the parts together into a new whole, stressing creative behavior, forming of new patterns, to produce a unique communication, a plan of operation, a set of abstract operations. The focus here is on being creative.

6. *Evaluation.* The ability to judge the value of material for a given purpose, judgments to be based on definite criteria. Bloom considers these to be the highest learning outcomes—an arguable notion in mathematics—because "they contain elements of all other categories, plus conscious value judgments based on clearly defined criteria." [1]

In a mathematics classroom, teachers provide students with all of these tools for learning, (1) through (6), but in our "own way," like the Red Queen, while the calculus student is receiving on Channel (1). Unfortunately for the teacher, learning cannot be taught: it needs to be achieved. What I find intriguing about Bloom's taxonomy coupled with the idea of writing assignments, is that we can, through writing assignments, achieve an applied, in-depth heuristic which we can have students follow to acquire an understanding of what it means to know something well. As a sample overnight exercise, students are asked to write about a concept following the steps below. The assignment could involve the meaning of the "Mean Value Theorem" or the definition of "limit."

1. *Knowledge.* How is the concept defined in the text? Write it out, as stated.

2. *Comprehension.* Write out what it means to you, in your own words. How does your statement seem to be different from that of the text? Could your notion of the concept be misunderstood? How can you be certain that the reader will know exactly what is meant? Is your notation accurate? What are you missing? Are you adding something?

3. *Application.* Specifically, in the next day or two, where will you find the concept useful? Or has it already been useful? In what way might it be important? Browse ahead in the text to answer this. Can you devise a problem that this concept can help us with? Are there any specific, concrete examples that you can now provide to supplement this concept or illustrate its point?

4. *Analysis.* What are the terms which this concept requires an understanding of? Write out all the terms involved in this concept, and their definitions. What terms appear to go together? What other results or concepts are related?

5. *Synthesis.* Write an introduction to this concept for addition to the text. Briefly sketch how the proof to this concept might go, or proofs which involve the concept. Invent a problem for which this concept would be useful.

6. *Evaluation.* If a grading scale is provided, how would you rate your own understanding of this concept? Would your writing above be publishable in this calculus text? If not, rewrite your explanation of the concept until it resembles that of the book. Why is this definition important in calculus per se? To you? Why should it be important to others?

That students will assimilate this method is too optimistic; however, the taxonomy is helpful to refer to when students fall short in understanding.

Examples of Writing Assignments

While the former heuristic might be a *model* for coming to know a concept through an individual assignment and while it even provides terms for *evaluating* an assignment, these categories also suggest ideas for the *design* of varieties of assignments. The assignments listed below are merely applications: I have not yet attempted to design them as a sequence as well. Obviously, providing a representative sample of student responses to all questions would be impossible; I have had to settle for providing merely a few examples of student writing and some evaluative remarks where the writing responses were short enough to include.

WRITING FOR KNOWLEDGE   Students frequently put a low value (as does Bloom) on remembering, or memorizing, but memorized definitions and theorems are the building blocks of mathematics. The student who says a series converges when the last term goes to zero is missing a fundamental understanding, as is the student who says that a derivative is when we find the slope of the tangent line, because we cannot do mathematics with definitions like these. These assignments force students to come to a state of understanding of what a definition requires, or at the very least, to learn to memorize.

Typical activities and assignments. Short, in-class definition writing. Brief, on-the-spot summaries of theorems or algorithms. Where are we now? What did we just prove? Where have we seen this concept before?

Characteristic Example. Define a polynomial in x over the Real Numbers.

Commentary. The results of this simple exercise never cease to amaze me. In this type of assignment student errors fall into something of a catalogue.

a. Is the student merely providing an example?

   * A polynomial is like $ax^2 + bx + c$.

   * $y = e^x x^x$ (which, of course, is not a polynomial.)

b. Is the student providing characteristics or qualities of the concept, explaining results which follow from the proper definition?

   * When you have a polynomial, you can factor, differentiate, and integrate it.

   * A polynomial is the equation of a curve.

c. Is the student merely telling how the concept is useful, giving its purpose?

   * We use polynomials in Taylor's polynomials to make a function simpler.

   * If $lim\, f(x)$ exists and $f(x)$ is a polynomial then $lim\, f(x) = f(x_0)$.

d. Is the definition given in terms of a process? (Warning signs are "when" and "-ing" words.) This appears to be a common problem—to some students, derivatives and limits are verbs, not nouns. This tendency of students to regard things as processes may reflect on their tendency to view mathematics as a rote activity.

   * A multiple addition and subtraction with different powers of x for each part.

   * An equation in which some constants are multiplied by variables which are sometimes raised to powers. These are then added and subtracted.

   * A polynomial is when you raise something to a power.

   * A constant with decreasing exponents and must exist.

e. Does the definition use the appropriate mathematical symbols?

   * $f(x) = a_x + a_x^{n-1}$ (This student remembers only that some notation is up, some is down.)

   * $a + a^2 + a^2 b + a^2 b^2$ (This student perhaps does not think of a polynomial as a function, so much as something to be factored.)

   * $a + bx + cx^2$ (This student has failed to grasp the need for "a_i"-notation.)

f. Is the definition too wordy? Or is the student basically settling for "damage control?" to maximize the partial credit? The student should provide only what is necessary and sufficient.

   * $f(x)$ is a polynomial if it is a number raised to some power which makes it a many numbered function. It is not a straight line.

   * more than one member of a term with the unknown variable (such as x) part of the whole term with the exponents of the

"x" term increasing as integers.
Example: Polynomial is $9x^3 + 6x^2 - 3x + 1$

* a polynomial is an equation where the variable has exponents that are always defined as the x^i and the a_i are never a fraction and whose function is continuous and smooth.

WRITING FOR COMPREHENSION   In mathematics we would like to test the level of remembering, to integrate remembered facts with other ideas, or problems. Drawing connections, translating from one mode to another, and summarizing are part of this level of understanding.

Activities, Assignments.   Rewrite a section of the text in your own words. Write a list of questions which you have on this section. How do you divide polynomials? How do the graphs of $y = x^2$ and $y = -1/x^2$ relate? State the limit theorems in your own words: where have you used them in solving problems? Summarize the different ways the definite integral has been used to date.

Characteristic Example.   Describe the graph of $y = 3x/(x-1)$ as if you were explaining it to a friend on the phone.

> The graph of this function looks like a hyperbola. The horizontal asymptote is at $y = 3$. The vertical asymptote is at $x = 1$. The graph goes from infinity just below the $y = 3$ line into the area of the vertex (which it crosses) then slopes downward and heads for negative infinity always approaching $x = 1$ but never touching it. The graph behaves a similar way in Quadrant I. It approaches infinity in both positive directions of x and y. The curve is always coming closer to, but never touches $x = 1$ or $y = 3$.

> The graph is a hyperbola with center (1,3). The asymptotes of the graph are $y = 3$ and $x = 1$ giving the graph a tilted look for a hyperbola. Being a hyperbola, the graph has two curves. The top curve approaches the line, $x = 1$, from the positive side (right) and follows the asymptotes to positive infinity (on the x-axis). As x becomes positively larger, the curve approaches the other asymptote, $y = 3$, from the top and goes on to positive infinity (on the x-axis.) The lower curve is reflected from the top curve with the line $y = -x + 4$ as the "mirror" or line of symmetry. This curve approaches the line $y = 3$ from the bottom as x becomes negatively larger. This curve passes through the origin. This is the graph's only x or y intercepts. As y gets negatively larger, the bottom graph approaches the asymptote $x = 1$ from the negative (left side).

Drawing connections, translating from one mode to another, and summarizing are part of ... understanding.

Commentary.   Since the objective here is to refine writing to the point of communicating, we can say that both of these attempts lack focus to the point of requiring a traffic controller. A teacher might well find it difficult to grade these, or to rewrite them. But the understanding of the reader is the objective for the writer, and we can show that simply comparing these two descriptions might be as instructive as revising. We might say to the class that the second attempt is probably clearer to everyone, and credit should be given that the student talks of symmetry, uses words such as "negatively larger," "approaches the asymptote ... from the top" and mentions intercepts. The first student fails more in vocabulary and uses vague expressions such as "the graph approaches infinity in both positive directions of x and y." The assignment, of course, was to describe the graph to a "friend" and many students will therefore tend not to use mathematically precise language; this reflects on how writing assignments can penetrate to the social dynamics of a classroom.

WRITING FOR APPLICATION   These days we look for applications in other fields to inspire students in mathematics, but finding applications may mean finding ways in which the subject becomes "real world" to the students. Of course, we are trying to frame a value system for them in which mathematics is real in itself.

Activities, Assignments.   Look up (some) concept in several sources—dictionary, encyclopedia, etc., and describe the differences you see. Explain how absolute value is used in "real life." Describe how to solve this problem as if you were writing to the boss of a company to which you were applying—be brief, and don't wing it, but be thorough. How does your calculator compute square roots? Describe how a slide rule works.

Characteristic Example.   Show in a few paragraphs how a knowledge of the improper integral can help us to understand series, and vice versa. Specifically work through some concrete examples we have talked about in this class.

Commentary.   While the responses would be too lengthy to reproduce, some common problems are easy to describe. In this assignment, most of my students presume I already know what they are talking about, and give themselves the opportunity to be vague. Here, a specific example is required, but most students made passing reference merely to the possibility of this. No drawings are provided, generally. Students apparently cannot see themselves as the authority in this assignment.

Most of my students presume I already know what they are talking about, and give themselves the opportunity to be vague.

WRITING FOR ANALYSIS   Students take apart a concept, and thereby their understanding is tested. Students must begin to draw deeper comparisons, see importance, organize.

Activities, Assignments.   Write a crib sheet on this topic for a friend. Describe the major results we have obtained on limits. Define all the terms used in the fundamental theorem of calculus. Explain how the definition of limit avoids the notion of a point "moving." In what ways can the definite integral be considered a function, and what are some of the properties of functions that we have studied which might pertain? Why is the differential so important? Rewrite p. 11 of the text, which we did not like, in the same style as the author, but improved.

Characteristic Example.   Write a two-page summary on Taylor polynomials and Taylor Series, as you would like to see such a summary at the end of the chapter. You should focus on organization, and include a maximum of mathematical information from these few chapters.

Commentary.   In this assignment, focus on topics is unevenly distributed, some students explaining in depth how to obtain a factorial, and dismissing series backhandedly. Some simply never do understand the theory. Nevertheless, students claim they find it exceptionally useful and they test on this chapter better when they do this exercise. This overnight assignment has counted 10 points toward the hour exam, and I can draw analogies between the assignments and the rest of their exams. The analogies are easy for a teacher to understand, but the students may be struck by them.

WRITING FOR SYNTHESIS   Creativity is a chief aim in mathematics, but we frequently withhold from the students a chance to show off their own creativity, and settle for their being able to answer correctly on a test. This exercise improves morale, and provides an opportunity to see some fun in mathematics. When students work in groups in creative activities, they go further, and are encouraged to think of mathematics as a collaborative subject.

Activities, Assignments.   Invent a problem for "real life" using the intermediate value theorem. Explain "infinity," how we have handled the notion in this class, and what you can and cannot do with the notion. Predict a theorem that's coming.

When students work in groups in creative activities, they go further, and are encouraged to think of mathematics as a collaborative subject.

Characteristic Example.   Refer to the graph on the board. From what you have seen of the derivative, can you predict on the basis of the slope of the tangent line, where a maximum and minimum will occur on an arbitrary curve? (If you can, you will have invented the "first derivative test.")

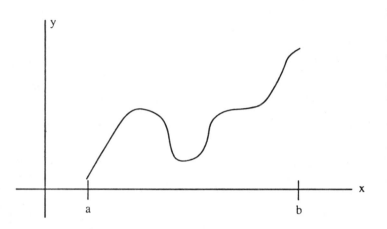

Derivative gives the slope of tangent line at a certain point. So if tangent slope is positive, graph is rising at that point. If tangent line is zero, graph is leveling off. At the leveling off, check the derivatives at the left and right of it, if the left is pos and right is neg, it's a rel max. Vice versa is a relative min. If they're the same—it says nothing.

When the slope of the tangent line is near or at zero, the curve at this point levels off. At that point, the minimum is found. As the slope of the tangent line increases the curve rises and reaches a maximum at a point. The derivative at this point is zero because it's not an endpoint. If the derivative is negative, this implies the curve is falling, etc.

Commentary.   While these students' responses are far from perfect, most students actually achieved a fair generalization, which would be true for any graph, not just that pictured. They responded with an "is that all?" expression, a teacher's joy.

WRITING FOR EVALUATION   This is the ultimate educational objective, where the student becomes the authority, the evaluator of his or her own understanding as well as an evaluator of the importance of various topics, and ultimately, the importance of the subject. Hopefully, the student, who has put the time and energy into learning can make these evaluations with care and thought. Group exercises are excellent anywhere, but I put the emphasis on them here because groups help students test their own understanding.

Activities, Assignments.   Group quizzes, group papers. Groups can compete, writing on transparencies, on many of the activities already suggested. Design a worksheet for helping a student evaluate the correct statement of a theorem: what are some common problems a student might run into? Write a test for this unit and

exchange your test with a friend. List five theorems in calculus in order of importance to you, defending why you think they are correctly ranked. What would you like most to remember about this term of calculus, to carry over to the next term, and why? Exchange papers with your neighbor; your neighbor should play the role of editor, and make changes in your (summary, paper, etc.) Rewrite your work to improve your previous exercise.

Commentary.   Group work is generally so much enjoyed, it is a wonderful way to begin writing assignments. When group work becomes a feature of a classroom, students begin to study together after class as well. Students in general are reluctant to be negative evaluators of others, however. At this level, I don't feel the teacher has much of a role as evaluator; presumably, the students themselves are moving into this role.

When Alice finally does become a queen, she is scolded by the Red Queen to speak only when spoken to. "But," observes Alice, "if everyone obeyed that rule—". Alice has finally found her confidence and with it , her ability to reason—qualities we want to nuture in our own students. The failure of our students to be sufficiently attracted to mathematics suggests that we must find new ways of teaching that nurture interest, develop learning skills, and that (for our sakes as mathematics teachers), create values which can hold their own against negative stereotyping about mathematics— stereotyping which implies it is a bleak, stressful, lonely subject. The essential problem in learning calculus, of course, remains the same as it has always been; the facts of the subject must be assimilated, and students must learn to respond and be productive; however the problems we face in today's classroom are new ones, and new solutions are in order. It is reassuring that, with writing assignments in the mathematics classroom, we can still place the responsibility of *value-acquiring* back on the individual, improve the classroom situation, and produce students who are more sensitive to learning, more sensible and more alive.

REFERENCES

1.   Benjamin S. Bloom (ed.), *Taxonomy of Educational Objectives: The Classification of Educational Goals, Handbook 1: Cognitive Domain*, Mckay, New York, 1956.

2.   Sandra Keith, "Exploring Mathematics in Writing," *The Role of Writing and Learning in Mathematics and the Sciences*, Bard College Institute of Writing and Thinking, February 1989.

3.   ——, "Using Writing to Teach Mathematics," *The Mathematics Teacher*, December 1989.

Writing in Mathematics: A Plethora of Possibilities

Timothy Sipka
Alma College, Alma, Michigan

Introduction

Teachers of mathematics from grade school to graduate school are reaching into their toolboxes and removing an old and dusty tool—the writing assignment. They are experimenting with this tool in a wide variety of ways, and for a variety of reasons. Some teachers, at institutions where a writing across the curriculum program exists, are interested in improving the writing skills of students. Others are using the tool as a means for allowing students to generate ideas, express concerns, and confess confusion. Many teachers, believing that writing and thinking are umbilically linked, are using the tool as a vehicle to improve the thinking/learning skills of students. Most, I suspect, are using the tool for a combination of reasons.

Many teachers, believing that writing and thinking are umbilically linked, are using the tool as a vehicle to improve the thinking/learning skills of students.

Bill Kennedy, a middle school teacher in California, has his students write letters to him explaining what they understand, what they don't understand, and what questions they're pondering; Joan Country-man, head of the math department at Germantown Friends School in Philadelphia, has her students do free writing at the beginning of every class; Allan Weinheimer, head of the math department at North Central High School in Indianapolis, has his students write their math autobiographies at the beginning of a term, and at the end of a term, requires his students to write letters to next year's class explaining what the course is all about; Coreen Mett, math professor at Radford University, requires her calculus students to keep a journal; David Smith, math professor at Duke University, periodically makes assignments such as "write a paragraph identifying the relationship between linear equations and lines in a plane;" and others are using the tool in more traditional ways such as assigning research papers, journal article summaries, and essay questions.

As these examples suggest, the writing assignment is an extremely versatile tool. Whether used informally or formally, it can enhance the learning of mathematics. This paper, admittedly not the definitive work on the topic, will consider many of the ways to use the tool, provide specific examples of writing assignments, and discuss the benefits of writing.

Types of Writing

All writing assignments fall into two basic categories: informal and formal. An informal writing assignment is one in which the content is king; the reader is primarily interested in viewing a hardcopy of the writer's thoughts on a particular topic or issue. The writing may contain mechanical errors, such as mispelled wurds, but the reader's main concern is the substance of what is said. Informal writing assignments can be completed during class and returned without a grade; they are especially useful, as Robert Gremore

remarks, "when the goal is to help students understand material they are seeing for the first time." A formal writing assignment, on the other hand, is one in which the reader is concerned with both the content and quality of the student's writing. In addition to evaluating the substance of what is said, the reader also evaluates the structure, organization, and mechanics of the writing. Usually, formal assignments are graded assignments that are completed outside of class.

The tree diagram given below displays the types of writing assignments considered in this paper, and provides an outline of the presentation.

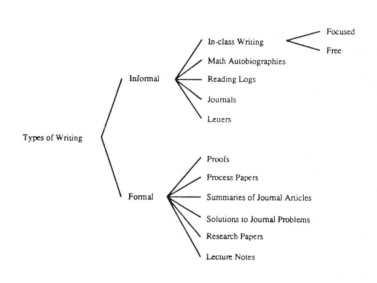

Informal Writing Assignments

IN-CLASS WRITING   In-class writing, as the name implies, is a type of informal writing that uses a small amount—ten to fifteen minutes—of class time. Students will spend this time writing about a particular topic you've chosen (focused writing) or one that they've chosen (free writing).

A focused writing assignment forces the student to discuss a particular topic that you feel is significant. You could ask students to identify the steps or heuristics used in solving a problem; you could ask them to summarize a lecture or film; you could ask them to critique the author's presentation of a topic; you could ask them to create problems or examples; you could ask them to explain their methods for solving a problem; and so on. The purpose of such writing exercises is to improve a student's understanding of the topic. "Students take conceptual ownership (of a topic) by articulating mathematical concepts in their own language," according to Paul Connolly, head of the Institute for Writing and Thinking at Bard College [24: 14].

In *The Craft of Teaching,* Kenneth Eble devotes an entire chapter to the topic of getting students to think. In that chapter, he emphasizes the important role that randomness plays in thinking, and he encourages the reader to develop ways to stimulate and capture this randomness [8: 32–33].Free writing, writing that proceeds without a plan, is one avenue available to teachers for capturing a student's random thoughts.

Joan Countryman discusses her experiences with free writing in *Writing to Learn* [**29**: 164–165]. At the beginning of every class, she tells her students, "You can write about anything as long as you're willing to have me see what it is." The students spend approximately ten minutes thinking and writing; although their writing doesn't have to be about math, about half the time it is. Later, she will talk with each student individually about what the student has written.

MATH AUTOBIOGRAPHIES   One of the most efficient ways to become acquainted with the students in your class is to require them to submit their math autobiographies early in the term. At our first meeting, I share my math autobiography with the class and ask them to submit theirs at our next meeting. My only request is that they recount at least one successful episode in their mathematical careers and at least one not-so-successful episode. Because the autobiographies are not graded, I make no demands regarding length and form. I do, however, suggest that they aim for a target of two pages and that they use a word processor.

I have experimented with this type of writing assignment in variety of freshman level courses: Introduction to College Mathematics, Precalculus, Discrete Mathematics, and Calculus. In all of the courses, I've observed two phenomena. First, students have vivid recollections of teachers, topics, and experiences. Second, the students seem to enjoy the assignment; they appear to be genuinely interested in letting me know about the good, the bad, and the ugly. They boast, they blame, they confess, they despair. They try to paint an accurate picture of the circumstances that have led to their current situations.

One student recently boasted, "I am the type of student who will sit down to do a problem set and not get up until I've finished all the problems and understood each of them." Another very articulate student confessed, "It's difficult for me to remember the exact date that my hatred for mathematics began, but the circumstances which led up to my numerous wars with the math world are still quite familiar to me today."

Requiring students to submit their math autobiographies can do more than simply acquaint you with your students. Conveying to your students that you're interested in each one individually, which can be accomplished through the autobiographies, is an excellent way to foster a positive rapport with your students. A positive rapport, as Joseph Lowman points out in [**17**: 15–16], can increase student motivation, enjoyment, and independent learning.

Conveying to your students that you're interested in each one individually, which can be accomplished through the autobiographies, is an excellent way to foster a positive rapport with your students.

READING LOGS   Reading a section of a mathematics text is a slow and arduous process for most students. The hybrid language of symbols and prose, the abstract nature of the content, and the carefully orchestrated sequence of deductions provide students with ample opportunities to become lost. Asking students to express their confusion in writing provides them with a comfortable and convenient way to alleviate the problems.

Included in all of my course syllabi is a request for students to compile a list of questions and concerns associated with their reading of the text. The students are given daily reading assignments and are expected to submit questions they have over the material. I collect their questions, answer them in writing, and promptly return them. Occasionally, when a student asks a penetrating question or makes an insightful observation, I'll share it with the entire class.

I don't expect all students to have confusion over every section in the text; therefore, questions are usually submitted on a voluntary basis. When discussing a particularly difficult section in the text—one I know will generate confusion—I'll require all students to participate.

JOURNALS   A journal is a collection of mathematical writings ranging from unfocused diary-like entries to focused (i.e., assigned) entries such as summaries of lectures and discussions of problems. It is probably the most widely recommended informal writing assignment: well over half of the references I've listed advocate the use and proclaim the benefits of journals. Because of its ubiquity in the current literature, there's no need for further discussion.

LETTERS   Asking a student to write a letter—to you, to the author of the text, or to another student—is one of the most effective and natural ways to use writing in the mathematics classroom. Giving students a specific audience and requiring the use of a familiar form of writing make the job of writing less burdensome. The personal nature of a letter to the instructor can also foster, like the math autobiography, a positive rapport with the students.

There are many types of letters you could ask your students to write. Here's a small sample of possible assignments.

1. Write a letter to your instructor identifying the topics in chapter X that are causing you the most trouble. If possible, explain why you think the topics are so perplexing to you.

2. Write a letter to your instructor expressing your feelings about your performance on the last exam. Evaluate your preparation for the exam, the instructor's effectiveness in presenting the material, and any other factor that could have contributed to your performance.

3. Write a letter to the author of the text identifying the good and bad points of the book (or a particular section of the book).

4. Write a letter to a student who plans to take discrete mathematics next term. Explain to the student what the course is all about.

    (At the beginning of next term, read a collection of these letters to the new recruits.)

Formal Writing Assignments

PROOFS   In his book, *Writing with a Word Processor*, William Zinsser makes the following comments about writing. "Writing is linear and sequential. Sentence B must follow sentence A and sentence C must follow B. Every sentence must be a logical sequel to the one that preceded it." [**30**: 102–103] We mathematicians have an opportunity to stress these characteristics of writing every time we assign a proof. In addition to its other virtues, a proof may very well be one of the best vehicles for emphasizing the linear and sequential characteristics of writing.

Assigning proofs also allows us to emphasize another important aspect of writing—rewriting. Requiring students to submit revisions of their proofs will give them a glimpse of what Zinsser calls "the essence of writing." [**30**: 26]

PROCESS PAPERS   A process paper is an expository paper in which the student is asked to discuss (i.e., explain, examine, critique, etc.) a particular topic—often a process. These formally written focused writing assignments are generally one to two pages in length, and students are required to write them on a word processor. Here are a few examples of past assignments.

1. You've often heard the claim, "division by zero is undefined." Explain why this is so. What are the mathematical reasons behind this rule?

2. Explain how one proves that a function is surjective.

3. Pretend you are the author of a calculus text. Write several paragraphs (or whatever it takes) introducing the notion of a limit. Be creative. Use diagrams, tables, etc.

4. Pretend you are the author of an encyclopedia of mathematics intended for mathematicians and nonmathematicians. With this in mind, write an appropriate entry for DERIVATIVE.

SUMMARIES OF JOURNAL ARTICLES   There are many well-written journal articles, accessible to our students and relevant to our courses, that are never read by students. Requiring students to summarize an article of this type can address this omission as well as provide students with an excellent way to demonstrate their comprehension of a particular topic.

Although a summary is one of the basic modes of college writing, it would be wrong to assume that all students know how to produce an accurate and clearly written summary. Therefore, before asking students to write one, I recommend that you distribute a handout (your English department may have one) that discusses the characteristics of a good summary—brevity, completeness, objectivity, etc.

SOLUTIONS TO JOURNAL PROBLEMS   This year, a colleague and I are offering a two-credit course called Problem Solving. Intended for mathematics majors, the course will focus not only on the finding of solutions to current journal problems, but on the writing of formal solutions. We hope that by writing for a real audience (an editor), the students will have increased motivation to write well.

There are many journals whose problem sections are accessible to undergraduates. Our course will use journals from the following set: *Pi Mu Epsilon Journal, The Pentagon* (journal of KME), *The College Mathematics Journal, Mathematics Magazine, Mathematical Gazette, Journal of Recreational Mathematics, New York State Mathematics Teachers' Journal,* and *Mathematics and Computer Education.*

RESEARCH PAPERS   The research paper, like the summary, is also one of the basic modes of college writing. However, for many of the courses we teach, it does not play the vital role that a summary plays. Assigned in the wrong course, the research paper could serve only to distract the students' attention from more important issues.

When I do assign a research paper, there are several steps I perform before the students submit their final drafts. Here are a few examples. For some other ideas, see [**5**: 42–44] and [**13**: 83–84].

1. Because it's important for students to find topics of interest to them, I initially give them a list of narrowly defined topics and allow them to choose topics from the list. If a student finds no topic of interest on my list, I help the student create one.

2. Before students begin the job of writing, I review the formats of several standard types of research papers: reports, compare/contrast papers, position papers, etc.

3. I have found it helpful to discuss the following issues at the outset of the assignment: plagiarism, when to quote, abuse of the passive voice, expressing thoughts in a clear and uncluttered manner, sentence length, etc.

4. I require students to submit rough drafts several weeks before the final drafts are due. This gives me an opportunity to provide comments about their writing, suggest changes, and consider the appropriateness of their sources.

LECTURE NOTES   The last type of formal writing assignment I'll mention is the preparation (by students) of a polished set of lecture notes. For each lecture, one student is the principal note-taker. It then becomes the principal note-taker's responsibility to provide a polished set of notes for that day's lecture. The student first submits the notes to the instructor; the instructor examines the notes and may request that revisions be made; finally, copies of the revised lecture notes are distributed or made available (placed on reserve in the library) to the entire class.

Benefits of Writing

Many benefits of writing can be accepted as axioms. For example, writing (especially in-class writing) assignments add variety to the typical lecture-oriented math class; writing assignments improve a student's writing skills; writing assignments help students become autonomous learners [**20**: 19]; and writing assignments provide an accurate assessment of a student's level of understanding/confusion.

Writing ... assignments add variety to the typical lecture-oriented math class; writing assignments improve a student's writing skills; writing assignments help students become autonomous learners ... and writing assignments provide an accurate assessment of a student's level of understanding/confusion.

The most important benefit of writing, but not at all obvious, is that writing assignments improve the thinking skills of our students. A proof that this is true doesn't exist, but there is strong support for this claim from the ranks of educators [**3**: 79], writers [**28**: 9], and futurists [**22**: 152]. In their book, *Reinventing the Corporation,* Naisbitt and Aburdene voice their support of the claim as well as anyone: "If our students' thinking skills have deteriorated badly—and we know they have—perhaps it is because their writing skills have grown equally slack. More and more educators have reached the same conclusion and come up with the same solution: Strengthen the writing curriculum as an avenue to sharpen thinking."

Conclusion

My main purpose for writing this paper was to acquaint you with the utility and versatility of a tool not often found on our job site. I've presented a dozen different types of writing assignments—hardly a plethora—that one could use to enhance the learning of mathematics. Furthermore, I've tried to give you some good reasons for using the various assignments. I have not claimed, and trust that I did not imply, that writing is a panacea for the problems our students have in learning mathematics. Writing is simply one tool we have in our toolboxes; it cannot be used in place of all other tools.

In his discussion of the myth that teaching is less mysterious than it is, Kenneth Eble reminds us of the pedagogical panaceas that periodically sweep through education. He warns that programs, such as writing across the curriculum, "adhered to too strictly and seized upon to the neglect of other equally important matters ... can belittle teaching by reducing its complex and often mysterious nature." [**8**: 26]

REFERENCES

1. E. S. Bell and R. N. Bell, "Writing and Mathematical Problem Solving: Arguments in Favor of Synthesis," *School Science and Mathematics,* 85 (1985): 210–21.

2. F. Bogel and K. Gottschalk (eds.), *Teaching Prose: A Guide for Writing Instructors,* Norton, New York, 1988.

3. Ernest Boyer, *College: The Undergraduate Experience in America,* Harper & Row, New York, 1987.

4. Marilyn Burns, "Helping Your Students Make Sense Out of Math," *Learning 88,* 16 (1988): 31–6.

5. Grace Burton, "Writing as a Way of Knowing in a Mathematics Education Class," *Arithmetic Teacher,* 33 (1985): 40–5.

6. David M. Davison and Daniel L. Pearce, "Using Writing Activities to Reinforce Mathematics Instruction," *Arithmetic Teacher,* 35 (1988): 42–5.

7. David M. Davison and Daniel L. Pearce, "Teacher Use of Writing in the Junior High Classroom," *School Science and Mathematics* 88 (1988): 6–15.

8. Kenneth Eble, *The Craft of Teaching,* Jossey-Bass Publishers, San Francisco, 1988.

9. Christine S. Evans, "Writing to Learn in Math," *Language Arts,* 61 (1984): 828–35.

10. Francis Fennel, and Richard Ammon, "Writing Techniques for Problem Solvers," *Arithmetic Teacher* 33 (1985): 24–25.

11. T. Fulwiler (ed.), *The Journal Book,* Boynton/Cook Publishers, Upper Montclair, New Jersey, 1987.

12. A. R. Gere (ed.), *Roots in the Sawdust: Writing to Learn Across the Disciplines,* National Council of Teachers of English, Urbana, Illinois, 1985.

13. Robert Gremore, "Designing Writing Assignments for Your Class," Unpublished paper, Praire Writing Project, Metropolitan State University, St. Paul, Minnesota, 1983.

14. Marvin L. Johnson, "Writing in Mathematics Classes: a Valuable Tool for Learning," *Mathematics Teacher* 76 (1983): 117–19.

15. Sandra Z. Keith, "Explorative Writing and Learning Mathematics," *Mathematics Teacher* 81 (1988): 714–19.

16. Bill Kennedy, "Writing Letters to Learn Math," *Learning* 13 (1985): 59–61.

17. Joseph Lowman, *Mastering the Techniques of Teaching,* Jossey-Bass Publishers, San Francisco, 1985.

18. William F. Lucas, "Let's Bring Writing and Reading All-the-way Across the Curriculum, While Introducing Some Twentieth Century Mathematics into the Schools," *Yearbook (Claremont Reading Conference)* (1985): 79–87.

19. Coreen L. Mett, "Writing as a Learning Device in Calculus," *Mathematics Teacher* 80 (1987): 534–37.

20. Liz McMillen, "Science and Math Professors are Assigning Writing Drills to Focus Students' Thinking," *The Chronicle of Higher Education* (January 22, 1986): 19–20.

21. Cynthia L. Nahrgang and Bruce T. Petersen, "Using Writing to Learn Mathematics," *Mathematics Teacher* 79 (1986): 461–65.

22. J. Naisbitt and P. Aburdene, *Reinventing the Corporation,* Warner, New York, 1985.

23. M. K. Paik and E. M. Norris, "Writing in Mathematics Education," *International Journal of Mathematical Education in Science and Technology* 15 (1984): 245–52.

24. Judith Axler Turner, "Math Professors Turn to Writing to Help Students Master Concepts of Calculus and Combinatorics," *The Chronicle of Higher Education* (February 15, 1989) 1, 14.

25. Diane Vukovich, "Ideas in Practice: Integrating Math and Writing Through the Math Journal," *Journal of Developmental Education* 9 (1985): 19–20.

26. Margaret Watson, "Writing has a Place in the Mathematics Class," *Mathematics Teacher* 73 (1980): 518–19.

27. A. Young and T. Fulwiler, *Writing Across the Disciplines,* Boynton/Cook Publishers, Upper Montclair, New Jersey, 1986.

28. William Zinsser, *Writing with a Word Processor,* Harper & Row, New York, 1983.

29. William Zinsser, *On Writing Well* third edition, Harper & Row, New York, 1985.

30. William Zinsser, *Writing to Learn,* Harper & Row, New York, 1988.

GETTING STARTED

A Reply to Questions from Mathematics Colleagues on Writing Across the Curriculum

Emelie A. Kenney
Siena College, Loudonville, New York

We have heard much in the last few years about the state of mathematics education in the United States: our younger students do not compare favorably with those of other nations in test after test, fewer and fewer students are choosing a mathematics major as undergraduates, and our graduate student body is becoming increasingly less populated by Americans. Many reasons have been given in an attempt to explain this situation, and many possible remedies offered to ameliorate it on each level of study—elementary, secondary, collegiate, and graduate. There is one method of instruction that may be used to great advantage on at least the first three levels, and that is the inclusion of writing as an integral part of course work.

The use of writing in the mathematics curriculum has gained favor among many instructors and curiosity among many others who do not currently ask students to write in their courses. Over thirty talks on the topic were given at a contributed paper session of the Mathematical Association of America (MAA) at the January, 1988, joint meetings of the national mathematics societies in Atlanta, Georgia—a session that was so well-attended that, at times, there was standing room only in the meeting room. In Phoenix at the meetings the following year, even more talks were given, and these seemed to be similarly well-received.

Of the mathematicians who are aware that some of their colleagues use writing in the mathematics curriculum, however, many are skeptical about using it themselves. Their skepticism is informed by a variety of concerns that must be addressed by anyone who hopes to be convincing about the merits of the writing-to-learn approach to teaching mathematics. Some may be unfamiliar with the principles of writing to learn, or they may believe that writing has no place in the teaching of mathematics and that it is inappropriate or makes for "soft science." Further, time is limited enough as it is to cover all the material in, say, a standard calculus sequence, and instructors believe that the inclusion of writing would further limit the time available. Others argue that their job would be that much more difficult because of the additional work involved. Many do not feel qualified to do a good job with a writing component, or they may simply be uninterested in learning more about process writing and writing to learn. Finally, they may not know how to go about including writing in the mathematics curriculum.

This paper is an attempt to respond to many of these concerns. A brief overview of the so-called Writing Across the Curriculum agenda is given, followed by a discussion of how these principles apply to the teaching of mathematics. Specific ideas for their implementation in mathematics courses are offered, and some fine points on issues of time and grading policies are addressed.

I. The Writing Across the Curriculum Agenda

We learn to write by writing, and we learn a given subject matter by writing, speaking, and thinking. Students may not do enough of any of these things in any of their classes; they definitely do not do enough writing in most of them, including their English classes. When Toby Fulwiler [1], one of the early proponents of the Writing Across the Curriculum (WAC) movement, was asked to distinguish among the terms "writing across the curriculum," "writing to learn," and "learning to write," he responded that WAC is a synthesis of writing and learning—that we do more learning by doing more writing and that we become better writers by being better learners. Indeed, it has been long agreed that language and thought are intimately related: thoughts become clearer as we seek to articulate them; the articulation of our thoughts is as incoherent as our thoughts. As we refine the articulation, we refine the ideas. In fact, proponents of WAC stress that we must not think of writing as "frozen language," but as continually subject to change, as a *process*. On this view, it makes sense to claim that all teachers are language teachers, since the principal way that knowledge is attained and communicated to others is by language, oral or written, no matter what the field of endeavor.

We learn to write by writing, and we learn a given subject matter by writing, speaking, and thinking.

Writing, of course, is the vehicle of language in which we are most interested here, especially as it affects and involves learning. In particular, writing promotes student-centered learning. Fulwiler writes that students think education happens *to* them, whereas they should regard education as something that they make happen *for* themselves [2]. Writing promotes student *ownership* of an idea primarily in the following ways: first, by writing, a student puts ideas into his or her *own* words; and second, through the process of writing, a student gradually makes an idea his own, makes it part of the architecture of his or her *own* knowledge.

To this end, WAC proponents encourage what is called collaborative learning, or learning and writing with others. Collaboration can take many forms, including joint projects, the sharing of writing in the classroom, the formation of study groups, and the compilation and organization of a week's lecture notes for distribution to classmates, among others. Working with others builds confidence. As well, it prepares students for professional activity, much of which is collaborative in nature, particularly among mathematicians and scientists.

II. The Application of WAC Principles to the Teaching of Mathematics

One can easily see, in a general way, that the ideas put forth by members of the WAC movement can be applied in mathematics courses. There are some reasons for including writing in our curricula, though, that are very specific to mathematics instruction. For one thing, writing and rewriting help students learn how to construct and write proofs. The construction and write-up of proofs is a significant stumbling block for many students, including mathematics majors and prospective majors. Students who begin their college days majoring in mathematics have switched majors once computation has been deemphasized and proving propositions stressed. There are texts [3] that a student can read, but there can be no doubt that reading about how to do proofs, or being shown how to do them, cannot replace actual attempts to prove propositions or to produce second attempts, amended versions, more elegant versions, and so forth, whether those proofs are created by oneself, with peers, or under the guidance of an experienced instructor. We do not often encourage several rewrites of a proof, but we should. Not only can rewriting an attempted proof improve a student's proof, but also it deepens the student's understanding of the proposition he or she proves, or, if the proposition is false, leads the student to suspect its falsehood and, ideally, to search for a counterexample.

In addition to encouraging students to write in order to learn, we may generate their interest in learning to write. Many mathematics students (and students of engineering and other disciplines) believe that they do not need to be proficient writers because of what they perceive to be the nature of their chosen field. On the contrary, mathematics or any technical language is a language, and ought to be used well. One *does* write in mathematics—mathematics is not simply strings of symbols, although student responses to examination questions requiring explanation sometimes makes it seem as if students believe that it is. Further, no prospective employee is going to get hired, no matter how dazzling the resumé, if his or her cover letter seems subliterate, because, finally, a lot of what professional mathematicians do is write: we write journal articles, lab and other technical reports, and grant proposals, among other things. Writing assignments for a mathematics course can be geared to the actual requirements of professional writing (more about this later).

Collaborative learning has a place in the mathematics curriculum because it affords the student practice in learning to write, an opportunity for writing to learn, and a chance to cooperate with others in ways that he or she will be called upon to cooperate later as a professional. Additionally, collaborative learning fosters trust in groups, which can aid in overcoming "math anxiety," as it is called. At the Centennial Celebration of the American Mathematical Society (AMS) in Providence in 1988, Paul Sally of the University of Chicago commented, "There's an abject fear of mathematics. It's common to hear, 'I was never any good at math.'" [4] Probably every mathematician has had individuals remark to us, upon discovery of our chosen profession, "Oh, I can't even balance my checkbook," or has heard similar comments, often delivered almost with pride, and sometimes with defensiveness, distaste, or a hint of sadness or regret. This kind of fear and dislike starts at an early age in many cases, but even in a student's college years, it is not too late to intercede. The group setting provides support and encouragement to students who lack confidence in their ability to do mathematics. Writing in groups fosters trust in the student's own ability to communicate mathematical ideas, and, as well, it gives crucial feedback to the student on just how much he or she has learned—something the student may not even have realized while concentrating on negative feelings and attitudes.

Writing in groups fosters trust in the student's own ability to communicate mathematical ideas.

Many of us have learned of and have been very interested in the success Uri Treisman has had using collaborative learning to motivate groups of his students at the University of California at Berkeley [5]. Treisman noticed in his mathematics classes that the Asian students outperformed the non-Asians and that the non-minority Americans outperformed the non-Asian minority students (Blacks and Hispanics, in this case). He wondered why the Asian students, who *are* minority students, outperformed all other students. He noted that the Asians form a community and engage, as members of that community, in studying, eating, rooming, and socializing. Other students, including those who also form communities, engage together in the last three activities, but not the first. Treisman hoped to convince other groups to mimic the ways his Asian students behave. He did convince a group of Black students to engage as a group in all of these activities, including studying, and then they outperformed their non-Asian counterparts. Later, he extended his observations to a system of minority mathematics education. While Treisman's experience offers simply anecdotal evidence, it does suggest the possibility that collaborative learning can make a difference in students' experience of themselves as learners of mathematics.

III. Some Ways to Implement Writing in the Mathematics Curriculum

There are myriad tools for incorporating writing into mathematics courses. Most of the following suggestions are geared to the undergraduate curriculum, although many can be used at other educational levels. One of the most commonly used, it seems, is the mathematics journal (also called by some a log or daybook). A mathematics journal is not a diary, although, certainly, it can contain some diary-like entries. The journal may include solved or partially solved homework problems, proofs, and details of proofs outlined in a text. It may also include attempts to explain an idea, definition, theorem, problem, or drawing. Pictures may be drawn in such a log, and it may contain responses to readings, examinations, class notes, and talks, as well as replies to philosophical or historical questions posed by the instructor.

We may also request our students to keep weekly abstracts, each entry containing a précis of that week's lectures and readings. In this way, a student can encourage himself or herself to keep up to date with the material covered in the course. As well, students gain a bit of exposure to a kind of writing they will be expected to do should they eventually give talks at national meetings, write articles for professional journals, or prepare doctoral dissertations, all of which require the submission of an abstract.

Other ideas for writing assignments include mathematical autobiography, letter writing to an instructor or text author, submission of solutions to problems in journals such as *Mathematics Magazine* and others oriented toward collegiate mathematics, and the preparation on a rotating basis of lecture notes for distribution to peers [6]. Students may be asked to summarize the points of a guest lecturer to the department, including prospective faculty, or of a professional journal article. They may create a table of contents for their journals or set of abstracts, an activity that can develop a considerable amount of organizational skills. There are always papers, although that may be too much to ask of first year calculus students, even if such a paper is simply expository in nature. Examination questions that call for explanation rather than computation may be a better alternative for these students.

This writer has enjoyed particularly heartening success with a special two-part assignment to students in an upper-level geometry course that combined an oral presentation and a writing assignment. The first part of the assignment was a presentation to the class of a journal article, perhaps one mentioned in the text (Coxeter's *Introduction to Geometry*) which dealt with material connected to that covered or touched upon in the course. The students were told that they should start thinking of themselves as young professionals. These presentations were treated the way we would treat a talk at a professional meeting: the speaker and the topic were introduced, the speaker used the board or overhead projector (or posters, in a couple of cases), or gave handouts, and the other members of the class took notes, asked questions, and applauded the speaker at the end of his or her presentation. In addition, each student and the instructor presented written comments to the speaker. These comments and suggestions were then incorporated into a formal write-up, satisfying certain standard publishing requirements, using a word processor, and following the usual stylistic guidelines. The talks and papers were almost uniformly fascinating. Some of the topics were taxicab geometry, almost congruent triangles, and the Brocard point. One student gave a detailed exposition of a section of a recently published article on the sequence of pedal triangles [7], and her performance and its reception served to replenish some of the confidence she had lost as a result of a previous unsuccessful experience in mathematics. A student who discussed connecting four points via a minimal route was pleased to note that minimal surfaces are a topic of current interest, and another student, who spoke on the four-color problem, was fortunate to discuss his topic with John Koch, a member of the department who had worked on the four-color problem with Haken and Appel. Most important, the students seemed to appreciate being taken as seriously as they were. Although the assignments involved a great deal of additional work for the students as well as the instructor, the outcome for the students was well worth the time and energy spent.

IV. Time Constraints, Grading Policies, and Other Concerns

Make no mistake about it: creating and implementing a writing intensive mathematics course requires many extra hours of work for the instructor, but there are steps one can take to minimize the amount of time necessary for a successful venture. Here are a few suggestions. First, one can cover fewer topics than one normally covers, and those that are treated can then be covered in greater depth, both by the instructor in his or her discussion of the topic and by the student in his or her writing. Calculus courses, for example, should not be survey courses, yet because of the sheer volume of material we attempt to present, they often seem that way to students. While the better students readily grasp whatever is presented to them and often even dig a little deeper on their own, some students are overwhelmed enough to drop a course or to be less successful than they might have been otherwise. We are not in a position to turn students off from collegiate mathematics so early in the game. We might recognize, instead, that some students require more explanation and more confidence building than others, and then give it to them. Responding to their writing efforts affords us an opportunity to do just that. (Of course, this assumes we have gotten them in the course in the first place.) The brighter students profit from response to their writing and thinking, as well, just as professionals benefit from revising their work, discussing it with, and receiving feedback on it from colleagues.

Further, we need not simply abandon material that is usually presented in a course; we can give students the primary responsibility for covering some of the material. This does not mean simply asking students to read a section or chapter of the text on their own. Rather, we can make it clear that we will be there for them outside of class should they need assistance in reading the text. Naturally, our offices will be slightly more crowded during office hours, but not taxingly so, and, besides, a crowd of students eager for help is not an unpleasant sight, particularly when one is accustomed to apathy and a "let's just get by" attitude. By the very nature of the enterprise, a student cannot maintain such an attitude when he or she has to have some part in teaching the material to himself or herself. Lest some object that this involves sloughing off one's instructional responsibility onto the student, let us note, first, that the instructor is actually more involved with teaching the material to students one-on-one. Second, after completion of their degrees, graduates will have to read texts and articles essentially on their own, and this experience offers them a gentle introduction to the practice while they are still students and can receive guidance. Finally, when students share the responsibility for their education and start to become less passive learners, they learn more and they become more interested in what they are learning. Passive reception of information does not encourage—and may even discourage— creativity and independent thinking [8].

Since a successful writing component requires that the students revise their work and that the instructor offers feedback, a good idea may be to abandon weekly or frequent quizzes in favor of, say, journals or abstracts, which may be collected every few weeks. In this way, students not only gain information about how they are progressing, but actually get the opportunity to correct errors, whereas they are usually not allowed to redo quizzes.

When students share the responsibility for their education and start to become less passive learners, they learn more and they become more interested in what they are learning.

Much of what has been said above, of course, is impractical for a large department in a large university, where lower level lectures can contain hundreds of students. Most, if not all classes, however, offer recitation or discussion sections run by teaching assistants, and grading is done by grading assistants. There is no reason why an assistant cannot give crucial feedback to students. The fact that so many assistants are foreign-born and do not use English as their primary language is not necessarily a detriment, since the fine points of English composition are not what is at issue in a writing to learn program: grammar and spelling are secondary to using writing as a tool to learn. Admittedly, it may not be possible for such individuals to help students learn to write. (It must be pointed out, however, that in many cases, those who learn English as a second language are more proficient than are some who learn it as their first.) In the upper-level courses, where enrollment runs more to thirty or so students, some of the WAC suggestions may, in fact, be practical, although they may not be preferable for university professors who need to spend a great deal of time on research and publication. Still, in a beginning complex variables course or linear algebra course, say, one can give fewer examination questions involving computation and more that require proof or explanation. Also, students may be requested to take turns compiling and distributing lecture notes; a university professor probably has the time to review a week's notes with the student who produced them that particular week. Students who work in groups on solutions to problems that appear in collegiate mathematics journals could reasonably expect response to and guidance about their work on, say, a weekly basis. In-class writing assignments are undoubtedly out of the question, not only for university classes, but probably for classes in a small college, as well, and may be suited more for remedial classes or classes in elementary and secondary schools.

Many instructors wonder about grading policies when considering including writing assignments in their courses. One hears, for example, from instructors who wonder how such assignments should be graded, for how much of the final grade these should count, and whether they should be graded at all. First of all, the student should probably be given credit for completing requirements of a course, so most instructors will want to grade writing assignments. Given that such assignments will most likely be graded, here are some ideas for grading policies with respect to, say, journals. One could take a standard route: give letter grades, and count as a percentage of the total grade. Or one could drop one letter grade (or half a letter grade) if writing is *not* done. Henry Steffens of the Department of History at the University of Vermont suggests simply not accepting an assignment until it is acceptable (thus encouraging the revision, rewriting, and editing process) [9]. Stephen BeMiller [10] of the Department of Mathematics at the University of California, Chico, recommends grading journals as follows: "Engaged, +2; required, +1; bypassed, 0."

Some instructors, having had little or no formal learning in creating writing assignments, wonder, too, just what makes a good writing assignment generally. This question was put to a group of educators at an interdisciplinary workshop on writing across the curriculum at Martha's Vineyard in July, 1988. Some of their responses included the following: connection to or application of material covered, student choice of topic or aspects of a topic, provision of opportunities for revision, availability of feedback and opportunity for response to feedback, provision of clearly written instructions, and the creation of an assignment that is not mere busy work.

V. Sample Journal Questions and a Sample Student Journal Entry

Devising writing assignments for mathematics courses, while it is a task that requires a certain amount of energy, is not beyond the ken of mathematics instructors. Some questions may be created in advance of presentation to the students, and some arise naturally during the course of instruction. What follows are some samples presented in a multivariate calculus course and others given in a course in differential equations.

The first set focuses on the issue of the student as learner. Here, he or she is asked to become conscious of *how* he or she learns mathematics, and the issue of mathematics anxiety is subtly addressed.

> Describe a problem or topic you found difficult. Analyze the difficulty. Try the problem again or read again about the topic. Analyze again. Repeat the process until you feel comfortable with the problem or topic.

> Describe a problem or topic you found easy. Analyze as in #1, above.

> Describe a problem or topic that you found particularly satisfying, fascinating, or beautiful. What made it so? Analyze your reaction. How is this problem or topic different from others? Similar to others?

> List all the mistakes you have made on homework problems, in class discussion, and exams. Do these errors have anything in common? Can you categorize them? How do you think you can learn from them? What do your particular errors tell you about the way you approach a problem?

Students *must* believe they are active participants in the academic arena if they are to avoid the passive receptor syndrome.

Students' opinions on how their education is progressing are seldom solicited, particularly in these days of the return to the "core curriculum," which largely ignores the issue of student empowerment. Students *must* believe they are active participants in the academic arena if they are to avoid the passive receptor syndrome, and there is no good reason why they should not question what is presented to them, Bloomites to the contrary. Engineering majors have a lot to say about the following topic, and what they have to say is worth a hearing:

> How would you alter the curriculum in your field and in mathematics? Would you delete mathematics courses? If so, which ones? Would you add any mathematics courses, and, if so, which ones? Which sections of your text would you like to see augmented? Which topics would you like to see covered in greater detail? Lesser detail? Be specific.

The next set of questions illustrates how journal entries can call for imagination, analysis, or generalization. The second question arose after an hour long discussion of smoothness in a multivariate calculus course.

> If you had to give an interpretation of n-space, for $n \geq 5$, what might it be? Use your imagination, and be detailed in your response.

> Analyze the definition of "smoothness" and the definition of "piecewise smoothness." Are they good definitions? What makes a definition good? Would you change these particular definitions? If so, how would you change them? What *is* smoothness? piecewise smoothness?

> Conjecture a generalization about the directional derivative and gradient definitions for n-space.

Another example of *ad hoc* questions arose as a result of a student query about the necessity of learning computational problems that are more easily solved with modern instruments such as calculators and computers. Students were asked to respond to that individual's question. Later, they were asked to consider the opposite side of the issue.

> Why do we learn how to do certain calculations, such as integration and figuring of a matrix's determinant, despite the availability of advanced calculators?

> Counter the argument that students must learn how to perform calculations by hand despite the availability of fancy calculators that do these calculations for them. In other words, argue that it should be unnecessary to devote time to learning a bunch of calculations when we have such calculators at our disposal.

In addition to responding to such broad questions, students from the multivariate calculus class solved some problems from their text in their journals. Here are two examples of student attempts at solving problems.

> Find the symmetric equations for the line that lies in the plane $y = 1$ and is tangent to the intersection of the plane and the paraboloid $z = x^2 + 16y^2$ at $(-3, 1, 25)$ [**11**: 745].

Solution [LB]:

> Any line that lies in the plane $y = 1$ will have "1" for their y coordinate. So we already know that one of the symmetric equations for the line is $y = 1$. Next take $f(x,y) = z = x^2 + 16y^2$. We know that $f_x(-3, 1, 25)$ is the slope of the line tangent to the curve, C, that is obtained when the graph of f and the plane $y = 1$ are intersected. Then vector $\mathbf{t} = \mathbf{i} + f_x(-3, 1, 25)\mathbf{k}$ is also tangent to C at $(-3, 1, 25)$. $f_x(-3, 1, 25) = 2x|_{(-3,1,25)} = -6$. So $\mathbf{t} = \mathbf{i} - 6k$.
> We will call the $\mathbf{i}$, $\mathbf{j}$, and $\mathbf{k}$ components of $\mathbf{t}$ a, b, and c, respectively, and $P(-3, 1, 25) = (x_0, y_0, z_0)$. The general symmetric equations are
> $(x - x_0)/a = (y - y_0)/b = (z - z_0)/c$.
> Substituting our values into this equation, we find
> $(x + 3)/1 = (z - 25)/-6$ and $y = 1$.
> The fact that we found $y = 1$ corresponds to what we said earlier about the y components of the lines in the plane $y = 1$.

The student who wrote the following two journal entries used the writing process to learn how to solve a problem from the text. One can see in her writing her thought processes, her hesitations, her moments of insight, and her satisfaction at having worked well on the problem. Her writing illustrates well some of the principles proposed by those interested in using writing to learn.

Suppose an airplane is flying in the xy plane with its body orientated at an angle of $\pi/6$ with respect to the positive x axis. If the air is moving parallel to the positive y axis at 20 miles per hour and the speed of the airplane with respect to the air is 300 miles per hour, what is the speed of the airplane with respect to the ground? (Hint: the velocity of the plane with respect to the ground is equal to the sum of the velocity of the plane with respect to the air and the velocity of the air with respect to the ground.) [11: 624]

First attempt at solution [WZ]:

[Draws a picture.] So that's the picture I come up with. Now I want to find the speed of the airplane with respect to the ground. The hint tells me that the velocity of the plane with respect to the ground is equal to the sum of the velocity of the plane with respect to the air and the velocity of the air with respect to the ground.

velocity of plane to ground = 300 mph + speed of air with respect to ground

I'm not really sure about this problem. I don't know what the "respect to's" mean. First I'm going to try to see how they got 300 and if I can do that then I can get the rest of the problem. So let me redraw the picture on a bigger scale. [Does so.] I must be missing something. Help! Please!!
Wait what if that first part of the hint was just referring to finding the norm of the vector. So then you get something like
$$= [sic]\sqrt{300^2 + 20^2 + 0} = \sqrt{90,000 + 400} = \sqrt{90,400} \approx 300.67.$$
I know I'm missing something but I don't know what it is.

Second attempt at solution:

What I should have done to make things more clear for myself is I should have labeled things like letting v_{pa} be the velocity of the plane with respect to the air, and let v_{ag} be the velocity of the air with respect to the ground. I should have also taken the hint more seriously which stated that the velocity of the plane with respect to the ground, v_{pg} is simply $v_{pa} + v_{ag}$. Next all I have to do is put $v_{pa} = \|v_{pa}\|(cos\theta i + sin\theta j)$. I got this from another problem [gives reference]. Then by what I did in my first attempt at this problem I get $v_{pa} = 300(cos\pi/6i + sin\pi/6j)$ which then I get $= 300(\sqrt{3}/2i + 1/2j)$ which is equivalent to $150\sqrt{3}i + 150j$. From before I found out that $v_{ag} = 20j$. Now to find $v_{pg} = v_{pa} + v_{ag} = 150\sqrt{3}i + 150j + 20j = 150\sqrt{3}i + 170j$. But I want the speed so all I have to do is find $\|v_{pg}\| = \sqrt{(150)^2 + (170)^2} = \sqrt{96400} \approx 310.483$.

No one who proposes the inclusion of writing in the mathematics curriculum expects an instant cure to whatever plagues American mathematics education. But those of us who do use writing assignments in our courses notice more of a sense of accomplishment in the students, greater involvement in their learning, and a deeper understanding of material presented than they had previously exhibited. These are the results that make the extra work worthwhile.

REFERENCES

1. Toby Fulwiler, personal conversation, Martha's Vineyard, Massachusetts, July 2, 1988.

2. ——, "Writing is Everybody's Business," *National Forum*, 45 (Fall, 1985): 22.

3. Daniel Solow, *How to Read and Do Proofs: An Introduction to Mathematical Thought Process*, John Wiley & Sons, Inc., New Jersey, 1982. See also the classic works by Pólya and Lakatos.

4. Quoted in *The Providence* (Rhode Island) *Journal*, August 8, 1988.

5. Uri Treisman, "Mathematics Workshop Revamped," *Undergraduate Mathematics Education Trends: News and Reports on Undergraduate Mathematics Education*, 1 (March, 1989).

6. Timothy A. Sipka, "Writing in Mathematics: A Plethora of Possibilities," Joint Mathematics Meetings, January 13, 1989.

7. John G. Kingston and John L. Synge, "The Sequence of Pedal Triangles," *The American Mathematical Monthly*, 95 (August-September, 1988): 609–620.

8. Toby Fulwiler, *Teaching With Writing*, Boynton and Cook, Ports-mouth, New Hampshire, 1987, pp. 7–8.

9. Henry Steffens, *Interdisciplinary Workshop: Writing, Thinking, Learning Across the Curriculum*, Martha's Vineyard, Massachusetts, July 10, 1988.

10. Stephen BeMiller, *Interdisciplinary Workshop: Writing, Thinking, Learning Across the Curriculum*, Martha's Vineyard, Massachusetts, July 10, 1988.

11. Robert Ellis and Denny Gulick, *Calculus with Analytic Geometry*, third edition, Harcourt Brace Jovanovich, New York, 1986.

Writing in Mathematics at Swarthmore: PDCs

Stephen B. Maurer
Swarthmore College, Swarthmore, Pennsylvania

Swarthmore College has a writing-across-the-curriculum program with an unusual emphasis, especially as it impacts math and science. One reason often given for having all departments participate in writing programs is that the problems of writing are universal. However, a principle of our program is that the problems of writing in various disciplines are *different*. Consequently, in each writing course, freshmen and sophomores are asked to write papers using the conventions of that discipline. In the Mathematics Department, we ask students to learn the proper use of such things as definitions, notation, numbering, displayed theorems, and displayed examples, and to use these conventions to write good *expository* mathematics.

To dwell on these "fine points" of mathematical writing may seem like putting icing on the cake while the diners are choking on the main dish. That is, many students have so much trouble with basic mathematical ideas that most freshman/sophomore mathematics writing projects, I gather, emphasize brief, informal writing with the simple goal of helping students to get those ideas straight. Isn't that a better emphasis? I will argue that the Swarthmore approach is workable and has merit. But first I will explain the Swarthmore college-wide program, indicate how the Mathematics Department has responded, and discuss what the greatest difficulties have been.

The College Program

The writing program was adopted by the faculty in late Spring, 1985, as part of a major review of our curriculum for the freshman/sophomore years. A new type of course, a Primary Distribution Course (the PDC of our title), was defined. Then students were required during their first two years to take three courses in each of the divisions of the college—natural science and engineering (including math), social science, and humanities. Finally, two of the three courses in each division were required to be PDCs, from different departments.

The definition of PDC had several parts, of which the following two are the most important for our purposes.

1. PDCs are to provide a broad view of the scope and methodology of the department's discipline "in such a way that both those students who continue in the field and those who do not can profit substantially from taking them." [Quoted material is directly from the faculty legislation.]

2. PDCs "should develop students' capacity for reading, analyzing, arguing and writing *within the framework of the discipline*." [Emphasis my own.]

Each department was enjoined to offer at least one PDC. This was particularly a challenge in the Natural Science Division. Traditionally, serious introductory courses in this division were highly technical—they emphasized techniques one had to learn before one could get the "broad view" in later courses. Also, these courses were already so full of material that it seemed next to impossible to find time for writing as well.

The Mathematics Department Response

Swarthmore courses last a semester. The core mathematics offerings for freshman and sophomores are a year of single-variable calculus (Math 5-6), a semester of discrete math (Math 9), a semester of linear algebra (Math 16) and a semester of multivariate calculus (Math 18). There are also honors versions of Math 16 and 18. For less well prepared students there is a semester of "basic math" (Math 3) followed by a semester of "calculus concepts" (Math 4). There are also a variety of introductory statistics courses, ranging from one in which computer software replaces even the most elementary algebra (Math 1) to one that requires a semester of calculus (Math 23). Finally, computer science is given by the Computer Science Program, but some of the courses, including the introductory course, are cross-listed in Mathematics. A dependency chart is shown in Figure 1. (The dotted arrow from Math 16 to Math 18 means that students are encouraged, but not required, to take 16 before 18.)

One reason often given for having all departments participate in writing programs is that the problems of writing are universal. However, a principle of our program is that the problems of writing in various disciplines are *different*.

The math courses that have become PDCs are Math 1, Math 4, and Math 9. The Department has not attempted to make any of the core calculus courses a PDC. Although there has been ferment here about making these courses lean and lively, through Spring 1989 calculus has remained rather traditional, and the changes for 1989–90 either involve computer algebra systems or allow students more flexibility in sequencing material; the changes do not make these courses less technical or open up time for attention to writing. (However, there is one section of Calculus I, taught by our Provost, that has been given as a PDC. Enrollees are warned that it will be hard to go on to Calculus II from this section, as many standard techniques are not covered.) Introductory Computer Science (CS 15 = Math 7) is also a PDC.

Typically a PDC has two papers. Usually, each is submitted *twice*, once for comments on the quality of writing, and then in final form for a grade. Many courses have a writing associate (WA), a student trained in analyzing writing who evalutes the first draft. When there is no WA, the faculty member evaluates both drafts (a lot of extra work, even with PDC classes restricted to 25 students).

I will now discuss typical paper topics in our PDCs. Since I have taught Calculus Concepts and Discrete Math, I will speak mostly about these.

In Calculus Concepts, the first paper topic in 1989 was "What is an explanation." "What is a justification" would have been a better title. Students were asked to compare the standards of justification in mathematics with another area of their choosing—another academic discipline or something from daily life. I gave this assignment after repeatedly finding that students misunderstood me when I asked them exam questions like "For n a positive integer, why is $(ab)^n = a^n b^n$?" They tend to answer "It's a rule" rather than explain how this rule follows by expanding out both sides using the definition of positive integer exponents. (Yet I had emphasized the explanation from the definition in class!) I thought that if students were forced to write a paper about explanations, this might bring home more clearly what the mathematical standards for explanations are. In any event, the first paper in this lower-level course was thus *about* mathematics instead of *in* mathematics.

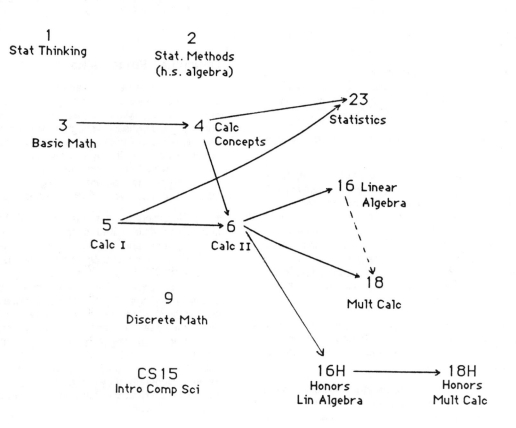

Figure 1

Flowchart of Freshman/Sophomore Math Courses at Swarthmore

For the second paper, students could take any major calculus topic covered in the course (the definition of derivative, curve sketching, max-min problems, etc.) and explain it to an imagined reader who had the prerequisites but had not seen that topic before. A five to ten page paper, in good mathematical style, was expected.

In Discrete Math, the first paper is usually short (around three pages). In 1988 the first writing assignment was:

> Write a paper explaining mathematical induction to a student who has had four years of high school math, not including induction. State the method of induction, give some examples, and convince the reader that the method is valid.

I usually assign the second paper after we finish a lengthy segment on graph theory. The assignment is:

> Introduce some applied problem whose solution is an application of an Euler graph theorem. Introduce the problem in applied terms, introduce only as much graph theory as is necessary to translate the problem into graph theory, prove the appropriate theorem, and translate the result back into a result about the application. Write the paper for a student who has not seen graph theory but otherwise has a background similar to your own. The applied problem can be taken from our text, or from reading in some other book, or can be something you think up yourself.

Thus, an important aspect of this paper is organization. All the theory students need is in the text and the lectures, but they need to cull out the theory they *don't* need. By "theory," I mean definitions and notation as much as theorems.

While the application topic was left to the student, notice that the mathematical content was limited to one of the Euler theorems. In some years I have let students choose the mathematical content freely as well. But this leads to a difficulty: students tend to get wrapped up in trying to learn some new mathematics (outside of the lectures) or they pick a problem for which the relevant theorems, though in the text, are quite hard to prove. Euler graph theorems are reasonable to ask our freshmen and sophomores to prove; a careful inductive proof of Dykstra's shortest path algorithm, say, is not. By specifying the paper topic, I allow the students to concentrate on the quality of exposition.

While originality in the application of an Euler theorem is not requested, occasionally I've gotten some delightfully original papers. A few years ago, a student solved a music harmony problem about creating a composition in which a set of chord progressions is performed with a minimum of repeats.

In addition to the main papers, there were occasional shorter writing assignments. Once or twice I evaluated a homework problem for writing. After students did particularly poorly on one test question in Math 4, I had them write a page recalling their thought process while they were answering that question. In Math 9, I once made transparencies of all the answers to a proof question on a test, projected them on a screen, and asked the class for comments on the writing. (Emphasizing the nature of mathematical writing won't fly unless you devote some class time to it as well as paper time.)

In Math 1 (Statistical Thinking), the first paper is on anything the student wants to say *about* statistics, for instance, how it gets used or misused in the newspapers. The second paper is *in* statistics. Each student must choose two variables, collect some data, analyze the level of correlation of the two variables, and write a statistician's report about it. In the main paper of the one PDC section

of Math 5, students are asked to give a thorough explanation of the limit concept as they understand it. (The exact nature of limits is emphasized in this section.) In Computer Science 15, students must write a computer program for a fairly complicated task and incorporate that program into a paper that provides thorough documentation.

Difficulties Students Have

As I tell students, writing mathematics is harder than ordinary writing. The arguments are usually more subtle and compact than in ordinary writing. Moreover, authors have to meet most of the standards of good ordinary writing and follow mathematical conventions as well. Not surprisingly, students have some writing problems that one would not see if they were writing ordinary prose, or if mathematicians were writing mathematical prose. (Students have plenty of ordinary writing problems too, e.g., their papers often lack a good introduction, but I won't discuss ordinary prose problems here.)

The main difficulty is that students rarely succeed at using mathematical conventions well, despite having seen them in math books for many years and despite my giving them written advice (see below). Here are several common problems: Students often do not see the need to use words in precise ways, nor do they take advantage of the convention that allows one to define words to take on precise meanings. They often try to explain general functional relationships in words, rather than write equations using $f(x)$. They don't make use of the opportunity to number equations and refer to them later by number (they used vague references instead). They tend to run their examples and results into their paragraphs, rather than break things up with displays of theorem statements and examples. Generally, they use a leisurely, discursive style that makes it hard for the reader to pin down the sort of specifics that make a good math paper.

Students often do not see the need to use words in precise ways, nor do they take advantage of the convention that allows one to define words to take on precise meanings. They often try to explain general functional relationships in words, rather than write equations using *f(x)*.

Guidebooks

The differences between math writing and other writing has to be pointed out to students. In addition to discussion in class, students deserve some written reference material on math writing. Unfortunately, there isn't much appropriate for them. Over the years, several excellent guides to mathematical writing have been produced [1], [2], and [3], but they are all written for mathematicians. In other words, they are written for people who already know the conventions for things like definitions, and know why it's valuable to have such conventions, but who may have trouble putting all the pieces together. College freshmen need to be told what the conventions are and why we have them.

I know of just one relevant document, unpublished notes by Steen [6] for students at St Olaf College. (Even these are written for math majors and undergraduate research students.) Consequently, I have written my own notes [4]. In fact, I have different versions for different courses (the more advanced the course, the more detail), and each year I make several changes, in light of additional faults I find in student papers. I would be pleased to learn from readers of other written material on math writing for beginning college students.

Is Our Writing Effort Worth It?

Two objections to the writing program described above are:

1. It is too ambitious, given the trouble most students have getting the simplest mathematical ideas straight.

2. It is irrelevant for the students it is aimed at, since most will not become mathematicians and will never have to write using mathematical conventions again.

I think both objections can be countered.

As for the first objection, Swarthmore is fortunate to have pretty strong students, whose general writing skills are usually fairly good even when their math is poor. So making Swarthmore students write math papers works pretty well. But Swarthmore is not unique in the quality of its students. Many other schools might have success with student papers. Certainly students find it more satisfying to write a paper than to write isolated paragraphs. Even so, here at Swarthmore we would probably benefit by having more writing in math class at the problem/paragraph level. When the PDC legislation was first before the faculty, some of us in mathematics suggested we would prefer requiring work at the paragraph level than at the paper level. But the sense of the faculty was that longer work should be expected, in order to make efforts at integration of material a part of the process.

As for the second objection, let me argue indirectly. When I was a high school junior, my English teacher announced the first day that, during the year, we would write stories in the style of Hemingway, Faulkner, Twain, and others. I was terrified. But I did my best and learned a lot from it. The teacher helped greatly: before he made us write in the style of Author X, we would read stories by X, and he would list salient aspects of X's style and give examples.

Once students learn that slight changes in wording or notation make tremendous differences in mathematics, they will be more on the lookout for those cases in ordinary writing where a small change also makes a big difference ...

I still remember these assignments because they helped me become very conscious of styles of writing and thus they better enabled me to fit each piece of my own writing to the need at hand. To my mind, teaching freshmen how to write mathematics has the same goal. For instance, learning about conventions for definitions makes students aware that words can have, and sometimes need to have, special meanings, and that when they do have special meanings the author needs to make this clear. Mathematicians aren't the only ones who need defined terms in their writing, although others (lawyers excepted, perhaps) are not so explicit about their definitions. For another example, the main purpose of the *statement* of lemmas, theorems, and corollaries within a paper (as opposed to the purpose of their *existence*) is to give a clear outline of the progression of the material. The fact that mathematics writing tends to be broken into short sections further helps with outlining the progression. Once students get the idea that mathematical writing can and should be clearly outlined, they can transfer this idea to their nonmathematical writing (I certainly have). For a final example, once students learn that slight changes in wording or notation make tremendous differences in mathematics, they will be more on the lookout for those cases in ordinary writing where a small change also makes a big difference (e.g., a mere comma determines whether a clause is restrictive or nonrestrictive).

Incidentally, slight changes in wording is one of the areas where good mathematical style is in conflict with good general style. When you have to refer to something repeatedly (your car, say), it is considered good general style to vary your vocabulary (write "auto" sometimes) to help keep your writing lively. But in mathematics, when you have given a word X a defined meaning, it is bad style to occasionally replace it with an ordinary-language synonym Y (unless you have explicitly defined Y to have the same mathematical meaning as X). Mathematical readers, used to small differences being crucial, will assume upon seeing Y that you *don't* mean X. If ordinary synonym Y has a mathematical definition too and it's different from X, substitution is even worse. For instance, if you have to refer repeatedly to a certain collection of things, in ordinary English it would be good to switch back and forth between "set" and "group." In a mathematics paper, doing this would be a guaranteed disaster!

Returning to my main points, I have argued that writing mathematics will help a person be a good general writer later. But we are trying to teach mathematics; is good general writing something we should worry about?

I feel strongly that the answer is yes. We mathematicians have always justified our undergraduate teaching in part by saying that we are teaching skills that are generally useful. When we say this, we are thinking of calculating and reasoning skills, but why not include writing skills too? Also, teaching the writing of mathematics does help teach the mathematics.

Final Remarks

I've made it sound like the PDC program is in a fully satisfactory, steady state. Not true. For instance, the stipulation that PDCs should be equally valuable for future majors and others is under debate. Second, the role of writing associates is being reviewed. Is it really possible for someone to comment on the quality of the writing and the argumentation without commenting on the content? But these issues are independent of the main point of this article: writing in the style of mathematics is a worthy activity to require of your beginning college students.

REFERENCES

1. American Mathematical Society, *A Manual for Authors of Mathematical Papers*, eighth edition, pamphlet, 1984.

2. Leonard Gillman, *Writing Mathematics Well*, The Mathematical Association of America, Washington, DC, 1987.

3. Paul R. Halmos, "How to Write Mathematics," in Steenrod, Halmos, Schiffer, and Dieudonné (eds.), *How to Write Mathematics*, American Mathematical Society, Providence, Rhode Island, 1973: 19–48.

4. Stephen B. Maurer, "Writing Mathematics Papers," unpublished notes, Swarthmore College; version for Discrete Math, 12 pages, revised November 1988; version for Calculus Concepts, 8 pages, revised, February 1990.

5. J. J. Price, "Learning Mathematics through Writing: Some Guidelines," *College Mathematics Journal* 20 (1989): 393-401.

6. Lynn A. Steen, Some Elementary Principles of Mathematical Exposition, unpublished notes, St Olaf College, May, 1973 (revision of 1968 notes).

A Writing Program and Its Lessons for Mathematicians

Ann K. Stehney
Center for Communications Research, Princeton, New Jersey

In the traditional freshman composition course of my day, the assignments involved literary analysis, an exercise having little in common with the thinking and writing I would subsequently do. While I learned and practiced the mechanics of good writing, the process was made more painful because I was struggling with the forms and conventions of argument in an unfamiliar discipline. On many campuses, the revolution of the 1960's brought an end to required courses like mine that had plagued generations of freshmen. Today's courses are likely to be less demanding, more "relevant," and voluntary. Undergraduates are now expected to recognize the weakness of their own writing skills and to seek remedies in elective courses in composition.

All told, I think I was better off with a tough, required, irrelevant course. But with writing across the curriculum, a tough required course need not be irrelevant. As the example of The Writing Program at Wellesley College shows, the benefits of this approach may extend to the entire academic enterprise.

In 1982, after fifteen years without a writing requirement at Wellesley, the two College committees responsible for matters of curriculum jointly proposed that all entering students be required to take a composition course. While about half of all freshmen enrolled in the existing English 100, the committees noted that there was no way of ensuring that the students who needed such a course took it. In addition, they explained, the issues of writing "pertain very broadly to a student's intellectual development, and it seems likely that all of our students—not just those whose writing is below college-level standards—could benefit from exposure to these issues. One of the goals of the proposed writing program is to make it possible for other courses to assume prior training in writing, but that is only possible if [it] is required." [1]

The original plan called for limited-enrollment courses for freshman specifically meant "to increase the competence of students in the writing of expository prose." [1] New or existing courses in various departments that were approved to fulfill the writing requirement would be identified by a common catalogue number, 125. The faculty teaching these courses would be encouraged to develop themes arising out of their own disciplines to provide direction and motivation for the writing exercises. It was hoped that the interdisciplinary dimension of the program would bring out a common concern for the relation between writing and inquiry in all fields of knowledge.

The English Department had long opposed requiring English 100 per se, partly because of a sense that instructors hired only for composition courses in the past had been regarded as second-class citizens. Now the department gave strong and crucial support to the plan, offering to contribute its faculty units from English 100 to the new program. The instructional cost of the program would be an additional 12 to 15 sections a year.

During debate on the proposed requirement, concern was voiced that course content might be sacrificed to the coverage of writing. The proposers of the program explained that, while the courses could in effect introduce students to particular fields in the curriculum, that was not their formal purpose. It was not intended that 125 courses use less challenging material than was found in other courses, but rather that somewhat less material might be covered in order to allow full attention to writing. Accordingly, the courses would not satisfy distribution requirements or requirements for the major in any department, nor would they serve as subject matter prerequisites for other courses.

To avoid such misunderstandings in the future and to minimize potential political problems, the proposal was revised to remove the 125 courses from specific departments and call them simply Writing 125. All sections of the course would be administered under a new umbrella, known as The Writing Program. The dean of the college and the head of the program, working in cooperation with department chairmen, would be responsible for staffing the course. The staff would come from the regular faculty, whose departments would be compensated with replacement faculty on a course-by-course basis. Overall enrollment patterns should be affected very little; one course for half of the freshman class constituted less than 2% of the college total.

Participating faculty would be offered training workshops on strategies and meet regularly to discuss their philosophies, goals, concerns, and experiences. The founding members would be asked to devise a set of practical goals for all sections and to adopt a common manual of grammar and style. Those who designed the program anticipated a great variety of approaches by the instructors, and they chose not to endorse any particular methodology. Nonetheless, they expected some consensus about effective methods to evolve from meetings of the staff.

The proposal called for considering the program a four-year experiment, to be evaluated at the end of the third year. With this understanding, the faculty approved the proposal by a large majority.

The program survived its 1986 evaluation and still looks much like the plan that was envisioned in 1982. Writing 125 is required in the first year of all entering students, including transfer students who have not completed a comparable course elsewhere. As reported in the Wellesley publication *Realia*, the philosophy remains that all students would benefit from "an introduction to academic discourse at a higher intellectual level than secondary schools provide." [2]

Incoming students receive a booklet describing The Writing Program and the topics available in Writing 125 in the coming year. They are told that "in all sections writing is taught as a means not only of expressing ideas but also of acquiring them. Students will receive instruction and practice in analysis and argument, in revision, and in the use and acknowledgment of sources." [3] They are asked to indicate their preferences from offerings with such titles as "Covering the News," "Renaissance Depictions of Gender," "Privacy and the Law," "Women in Fiction," "Language and Representation," "New Music," "Two on the Aisle," "Fairy Tales," "Whodunit," and "High Culture, Pop Art." An astronomy professor has dealt with science fiction, an anthropologist with the distinctions between magic and science.

"Writing is taught as a means not only of expressing ideas but also of acquiring them ..."

In its first three years, 37 instructors in 22 departments taught in the program. In 1988–89, faculty came from the Departments of Anthropology, Art, English, Greek and Latin, Music, Philosophy, Religion, and Russian. Members of various science departments have participated in past years, and the first mathematician has been recruited for 1989–90. The Mathematics Department had not been represented since I was involved in the initial planning of the program.

Marcia Stubbs, a faculty member experienced in teaching expository writing, was the first Coordinator of The Writing Program. Discussing the staff's assumptions, goals, and practices as they evolved, she explained that the program seeks to teach that writing is "a critical learning tool for students now and for the rest of their professional and personal lives."[2] As they see Wellesley's program, it should help instructors who want to include more ambitious written assignments in other courses. "One of the long-range goals of the writing program," she remarked, "is to increase the amount of writing required and the quality of writing instruction offered across the curriculum. A writing requirement allows the faculty to assume writing skills at some level. It deprives faculty of rationalizations for either failing to assign writing or failing to criticize the writing they receive."

"No matter what the topic or who teaches it, all the courses in Writing 125 share some common ground," according to Stubbs. "They are all devoted to teaching expository writing that depends on analysis and argument. We also try to teach students to behave like real writers. Writers, for example, may start out with nothing more than some scribbled notes that get reworked, then reworked again. A piece of writing goes through many stages—including stages in which you may take something barely readable and ask your best friend to read it. Or you take something a little clearer and ask some disinterested person to read it. Then you may take something that is said as well as you can possibly say it and send it off to some journal. When you get it back it's filled with the editor's suggestions for revising it. We try to encourage that kind of openness about sharing work with other people, and we try to teach our students to be good readers of other people's writing."

A writing requirement allows the faculty to assume writing skills at some level. It deprives faculty of rationalizations for either failing to assign writing or failing to criticize the writing they receive.

Students are advised and helped in various ways. Certain sections of 125 are recommended for those with little writing experience or for whom English is a second language, and a student with special difficulties can choose to take more than one 125 course. There is even an option for a student to take the course as an individual tutorial with a theme of her own choice; the tutor is an upperclassman, supervised by the Coordinator. Trained tutors and peer advisors are available, through programs that predate The Writing Program, to students in Writing 125 as in other courses. Librarians offer workshops to students on library skills.

No provision is made for advanced placement or exemption from Writing 125. (It is possible, as is the case with all requirements, for a student to petition a review board for an exception.) Stubbs spoke about advising students with top scores on the English AP exam who asked to be excused from the requirement. "It is true that a few first year students can write wonderful essays analyzing plot, theme, character, imagery, and symbolism in a play, short story, or poem. For them, I suggest that wonderful course on environmental ethics, for example. Good writing is good writing, but each discipline has its own forms, resources, conventions. This program gives students a way of entering into the thinking of many disciplines."

Stubbs described the basis of The Writing Program as follows: "The underlying assumption of the program is that writing is not simply a way of expressing or displaying what one has learned. Writing is itself a fundamental mode of learning. We learn from having written . . . We become critics of our own thinking processes as well as of our writing. We can see if our feelings and intuition can, in fact, be supported by evidence and if our conclusions are clear and persuasive to someone else. We may write for ourselves in order to learn, to understand, but paradoxically to write clearly we must write not only for ourselves, but for others who are not present, not known to us, and perhaps not yet born. That is why we have a Writing Program at Wellesley. It's why the writing lessons learned and the writing habits acquired in the writing courses should be repeated and reinforced across the curriculum."

The underlying assumption of the program is that writing is not simply a way of expressing or displaying what one has learned. Writing is itself a fundamental mode of learning.

The overall strategy of The Writing Program at Wellesley College is to teach students what writers do. Speaking to the Wellesley faculty in 1983, Elaine Maimon of Beaver College in Pennsylvania suggested that the first task is to change students' perception from "Writers publish" to "Writers go public with their work." Writers revise, she said; they reshape and reform their work according to the occasion and the audience. Writers know how to read their own early drafts critically, to submit themselves to questioning; students must learn to internalize questions that writers ask themselves. Writers accept and give help responsibly and tactfully; students must learn how to ask for advice and when to reject it. Writers make choices; they do not finish everything they start or include every thought or phrase that was committed to paper or go public with everything they finish. Only in school do people write under time pressure, then submit their first draft to scrutiny.

To be practical, here are some ideas and questions that I encountered when thinking about a 125 course of my own. Many of them were borrowed from or inspired by remarks of Marcia Stubbs or Elaine Maimon, but they have all been tailored to my concept of a writing-intensive course for students interested in mathematics. You may find them relevant or adaptable to assigning more expository writing in existing mathematics courses.

Advance Preparation

Read the literature on writing across the curriculum. Some student handbooks come with an instructor's manual and suggested syllabi; choose one that answers your questions and gets you started. I found good ideas in *Barnet & Stubbs's Practical Guide to Writing* [4] and Elaine Maimon's *Writing in the Arts and Sciences* [5].

Determine your priorities. Will you have time for both Writing to Learn and Learning to Write? Will you focus on generating, developing, and organizing ideas? To what extent are you willing to attend to problems of writing correct English, the source of first impressions? Your answer may depend on what help with writing is available to the students outside your course.

Choose your strategies. If you want to stress writing as a process, students might be asked to prepare written evaluations of their own work at various stages and to keep their drafts in a folder to be submitted with the final version. If you want to promote collaborative learning, they might be expected to read the work of their peers and complete written reviews by answering questions that you provide.

Get professional guidance on commenting constructively on your students' work-in-progress. Get advice on helping them review their classmates' writing with tact.

Pick a theme that lends itself to reading and writing assignments: the history of mathematics, the nature of mathematics, mathematicians talk about their work, etc.

Collect readings to show what mathematicians write. This should include published and unpublished (but polished) pieces, a variety of occasions (genres), written for various audiences. You might include biography and history, essays, review articles, exposition for experts and non-experts. Many items from publications of the Mathematical Association of America would be appropriate. Look also at articles, reviews, and letters in *Science, Scientific American, American Scientist,* and more popular periodicals. For unpublished pieces, consider sharing your correspondence, asking colleagues for their unpublished work, and using reports and other in-house documents.

Collect readings to show how mathematicians write. Include various styles and levels of difficulty depending on the purpose and the intended audience. Include pieces that you think are examples of poor exposition as well as good. Choose things that will lead to discussion, inviting the students to compare and contrast.

Collect readings that reveal the process of writing. Keep (and ask your colleagues to keep) drafts of a work at various stages. This is especially easy if you write on a computer or word processor. My own collection shows how a piece can change tone or focus from one draft to the next. My favorite one is a revised and polished letter-to-the-editor in which my main point had become buried; the version I eventually submitted looked nothing like this earlier "final draft." Early in the term, you might ask students to comment on a rough draft of your own.

Collect reference materials. Read (and require your students to read) the excellent manual by Gillman [6], the essays of Steenrod, Halmos, Schiffer, and Dieudonné [7], and Flanders' brief guide for hopeful *Monthly* authors [8] with its priceless examples. Be aware of good manuals of style, general resources for scholars such as Van Leunen's delightful *Handbook* [9], and the AMS's advice for preparing mathematics manuscripts for publication [10] and [11]. See Gillman's annotated bibliography for additional references.

Keeping your goals and chosen strategies in mind, plan exercises, assignments, and the use of class time. Expect to revise your plans during the semester.

During the Semester

State your policies clearly at the beginning; they may be quite different from your policies in other courses. Discuss the students' obligations in the course to proofread, to revise their own work, and to read and perhaps review other students' work. Cover such things as attendance, conferences with you, assignments (writing and reading), deadlines, and grading. For example, will you limit the number of absences or allow work to be made up? Do you require students to confer with you regularly outside of class? Will you grade on a pass/fail basis? Will the students have to absorb the costs of photocopying their work for peer review? What is your policy on word processors and the form in which work is to be submitted? Can students draft and review papers assigned by another instructor during this course?

What activities will take place in the classroom? Will you lecture or have discussions related to reading assignments? Will any class time be devoted to vocabulary, grammar, and usage? Will you ask the class to criticize or improve examples of writing that you provide? Will students write short exercises that can be shared and discussed? Will students work together or read their work to the whole class? Will time be devoted to peer review, either privately or in groups, or short conferences with you? Which discussions will take place in small groups and which will involve everyone? Remember that you must provide the structure for student interactions.

Written assignments may be frequent and short, or lead to the composition of a longer paper, or both. How many completed pieces will be graded? Must students submit their drafts, self-evaluations, or peer evaluations along the way and/or with the final version? Is a formal acknowledgment page required (to thank those who read drafts, for example)? Will topics be assigned? If not, must they relate to the readings?

Adopt conventions and a common set of terms and symbols for everyone's use in commenting on writing. Students should use these as they become readers of their own and others' work; it gives them a list of things to attend to (right word? tone? tense?). If they are to review each other's work, you should provide adequate guidelines for constructive criticism. You might include a checklist for evaluating the paper as a whole (organization, level, appropriate length, thoroughness in treating the topic, adequate introduction and conclusion, consistent voice, clear presentation of data, adequate references to previous work, etc.).

In reading students' work, withhold judgment on drafts. Comment on the process by raising questions and offering procedural advice; do not become a proofreader, editor, or coauthor.

Be Sure to Take Time to:

Discuss the philosophy behind the course and establish your credentials. This is probably more important in mathematics than other fields. Make it clear that writing tasks are part of professional life (well-chosen reading assignments may take care of this). Discuss the importance of writing as a tool of learning and inquiry and the role that writing may play within a process of intellectual development (look to essays on the subject). Make it clear that writing skills and habits need constant maintenance (this is harder—perhaps role models and anecdotes are required).

Articulate your goals and standards, such as to organize evidence in constructing an argument; to revise and edit as a normal part of the writer's work; to become useful readers for each other; to use resources effectively; to understand the conventions governing the acknowledgment of sources; to speak to an intended audience; to limit the topic as appropriate.

Discuss the process. Use your own experience. If you are an active scholar, you know what writing entails. Figure out what you do when you write, how you learned it, then teach your students to follow something like the same process.

Ideas for Writing Assignments

You might require that three or four revised and polished pieces of a suggested length (say, 4 to 6 pages) be submitted for grading during the term (say, after weeks 4, 8, and 12, or two each after weeks 6 and 12). Provide frequent exercises, such as the examples below, as the possible starting point for a finished piece. Allow students to choose which pieces they wish to work on further and eventually to submit. Withhold judgment until the students have "gone public" with their work: have them tell you when they think they are done, and avoid grading until then. Insist that they proofread their writing and read it aloud to themselves before making it public.

Some possible exercises for in-class or at-home writing:

Write a paragraph whose topic sentence is . . .

Summarize a reading assignment or discussion.

Analyze a reading assignment or discussion.

Write a paragraph about things on a given list or using a given list of words.

Revise a given paragraph to adapt it for a different audience.

Revise a given paragraph to remove jargon or make it more readable to the intended audience.

Do the opposite: translate an example of easily-understood prose into math-ese. (The result should be grammatically correct and adequate as technical writing.)

Write a paragraph in the style of *Time, Science,* a newspaper editorial, a book for 10-year-olds, a calculus textbook, a research journal, or whatever. (For this exercise, the content may be nonsense!)

Write a piece of self-referential prose, such as a journal entry, a job application, or a short biography.

Write a letter of recommendation, a letter to the editor of a publication, or a letter to a former teacher.

Review a book or article for a specific audience. Include both summary and analysis (original evaluation, criticism, or commentary).

Write an administrative-type document, such as a proposal for a new project, an evaluation of an existing project with recommendations, or a report based on an interview, a case study, or responses to a questionnaire.

Write a report for non-experts involving material from different sources, with proper attribution. This is the traditional term paper assignment, but the emphasis this time will be on the process of writing as much as on creditable coverage of the topic.

Write a technical report. It may be an expository report based on a section of a book or a review article, or it may report the results of an experiment, say with computer-generated data. The level of the paper will depend on your own background, but make clear who is your intended audience and what you expect the readers to know.

Write an essay in which you develop and present your own ideas or opinions. (Possible areas for a topic: the role of axiomatics and problem-solving in mathematics, the relationship between pure and applied mathematics or between mathematics and computer science, what is truth in mathematics, mathematics as an art or a science, etc.) It may be based on a single source such as Hofstadter's *Gödel, Escher, Bach*, Pólya's *How to Solve It*, or an essay in Newman's *The World of Mathematics*, but it must go beyond the ideas there.

You may notice that writing in the style of a research journal appears only as a minor exercise. I regard it as a specialized skill that is poor practice for other writing. While writing for a research journal may help a mathematician be completely rigorous, addressing a knowledgeable non-expert is more likely to deepen his understanding of the same work.

What might you accomplish in a course? Even in Wellesley's Writing 125, there is not time for students to become accomplished stylists. You can hope only that they develop the habits of a writer and acquire strategies for coping with their deficiencies. You can help by lowering the risks as you raise expectations for their writing.

ACKNOWLEDGMENTS. It is a pleasure to acknowledge the inspiration, encouragement, and down-to-earth advice I received from Marcia Stubbs on the teaching of writing. I am also indebted to Nancy DuVergne Smith, whose *Realia* article provided the quotes of Stubbs on the program that she so carefully nurtured. Stubbs and Kathryn Lynch are currently Codirectors of The Writing Program. For more information, the reader may contact them at Wellesley College, Wellesley, Massachusetts 02181.

REFERENCES

1. Memoranda to the Wellesley faculty (Academic Council) from the Committee on Curriculum and Instruction and the Committee on Educational Research and Development, dated November 17, 1982 and December 7, 1982. These memos document the proposal creating Writing 125.

2. Nancy DuVergne Smith, "Writing is 'in' at Wellesley," *Realia*, Wellesley College, May 1986.

3. Student booklets for The Writing Program, Wellesley College, 1988–89.

4. Sylvan Barnet and Marcia Stubbs, *Barnet & Stubbs's Practical Guide to Writing*, Little Brown, Boston, 1983 (teacher's manual available).

5. Elaine P. Maimon, et al., *Writing in the Arts and Sciences*, Little Brown, Boston, 1981 (teacher's edition available).

6. Leonard Gillman, *Writing Mathematics Well*, The Mathematical Association of America, Washington, DC, 1987.

7. Norman E. Steenrod, et al., *How to Write Mathematics*, The American Mathematical Society, Providence, Rhode Island 1973.

8. Harley Flanders, "Manual for Monthly Authors," *American Mathematical Monthly* 78 (1971): 1–10.

9. Mary-Claire Van Leunen, *A Handbook for Scholars*, Alfred A. Knopf, New York, 1979.

10. *A Manual for Authors of Mathematics Papers*, eighth edition, American Mathematical Society, Providence, 1984.

11. Ellen Swanson, *Mathematics into Type*, American Mathematical Society, Providence, Rhode Island, 1979.

Writing in the Math Classroom; Math in the Writing Class or, How I Spent My Summer Vacation

Thomas W. Rishel
Cornell University, Ithaca, New York

In 1973, at Dalhousie University, I was asked to teach a beginning statistics course with a calculus prerequisite. In place of a final examination I decided to assign a final project. The project was to be about one of two broad topics: the student's major, or "something having to do with Canada." Additionally, I told the students that they "must use the techniques of the course."

Everything went well until I received the final papers. I had no idea how to grade them, either "in the large," or "in the small." I spent an entire weekend merely reading papers, afraid to put a mark on them for fear I would appear foolish. Finally, on Monday morning, with grades due the next day, I began to see a strategy evolving for me.

Before I define my "strategies," however, let me mention one paper that had a great deal to do with shaping them. This paper was written by a Catholic nun. She went back to her home area in Newfoundland to survey people who had been moved from the "outports," those areas of Newfoundland nearly inaccessible by land much of the year. The government of Canada had assured such people that life would be vastly improved if they were to move to the towns where they could receive support all year round. In the meantime, the government had also informed these settlers that, if they did not move, they would no longer receive postal and other government services.

The nun's survey showed that almost all the participants agreed that "things had got better" for them since they moved. But then, however, she reported the anecdotal comments written on her survey forms or spoken to her during discussions. In every case, people revealed a sense of loss or ambivalence toward what had been done to them. They felt that the marginally increased services they had been given by the government could in no way make up for the loss of the sense of family and community they felt so strongly.

The project ended with the author's discussion of how a survey can give data which is both "correct," and yet at the same time be completely oblivious to the deeper questions involved. With this paper, the nun had cast into doubt a number of ideas: What does the design of a survey have to do with its results? Will the authority of the questioner skew the answers? To what extent can we quantify feelings? What is the role of govenment in social policy?

Writing across the curriculum, as it is generally called, has grown remarkably in the past few years. A real interest has now sprung up in having students use writing, just as they would use a computer or a library, as one of their tools to a fuller understanding of the content of the various disciplines they may study. As this movement continues, more and more faculty will be asked to get involved in the giving and grading of writing assignments in the classroom, and it behooves us to do some advanced planning before taking on this task.

To you mathematicians, let me state that whatever your reason for interest in the topic, I would like to point out an underlying fact. The state of mathematics is such that there is a significant decline in the numbers of students pursuing it, and an alarming decline in those claiming to "understand" any of it, and any attempt to popularize mathematics, whether in the general community, or among the target group of those who may choose it as a career, is to be encouraged.

Some of what I now tell you is "revisionist history;" under no circumstances could I, back in Nova Scotia, have enunciated much, if any, of what I am about to say; yet a lot of it was there in the grading that I did of those early projects.

The state of mathematics is such that there is a significant decline in the numbers of students pursuing it, and an alarming decline in those claiming to "understand" any of it, and any attempt to popularize mathematics, whether in the general community, or among the target group of those who may choose it as a career, is to be encouraged.

First, let me point out that there is a difference between:

1.   a good project,

2.   good exposition, and

3.   good mathematics.

A *good project* uses the course itself; it meets the criteria set out by the instructor and the text; it asks a "meaningful question" within the boundaries set out by the students' and course's level. It may fully explain the solution to the question, or it may raise the consciousness of the student by leading to a more delicate question.

Good exposition should set out that good question in an interesting and clear manner. It should exhibit the methodology carefully, while justifying the underlying assumptions, and indicate areas where further work is needed.

Good mathematics is at or above the level of the course; or, if it is "below" that level, it is at least justified by the nature of the project. By that I mean that the project itself must somehow ask a "deep question."

I mention all this, not because the above is a complete set of criteria for a valid project, but because the students will be asking you questions about your assignment, and you will at some point need to have thought a bit about what your project should measure. It is not sufficient to say, for a final project, or, for that matter, any other assignment, "We will have a final project in the course; make it ten pages typewritten," and then throw the topic away.

I note some points here.

1.   Instructor interaction far prior to the final report is essential.

2.   Students will have questions of the type:

    a.   Is my project acceptable?

    b.   Where are useful materials?

    c.   Can I share information?

    d.   What do you, the instructor, want?

Much of what I have just said relates back to the question of "How to grade—in the large;" but I am getting far ahead of my story, so let me return to my chronology.

It was a long time before I again felt brave enough to try some "alternate methodology" in one of my math classes. In 1985, however, I signed on to Anil Nerode's Exxon grant at Cornell University to design a geometry course for students who had no college mathematics. I knew that I needed to have a course that dealt with the fundamental concepts a geometer deals with; I also knew at the same time that I must consider my audience very carefully in designing the course.

Along these lines, I wanted to talk about, for instance, Euclid versus Noneuclid (that famous, forgotten geometer); straight versus curves; geometry and topology; surfaces; spacetime and four-dimensional space. I especially wanted to lead the students to such questions as: from where do these concepts arise; and, do they have "validity," or are they mental constructs. Meanwhile, I needed to work from "what everyone knows about geometry," namely, high school math (whatever that is); Pythagoras; distance and measurement; and so forth.

In the main, I wanted to show that "what one knows" is often "what one assumes," and I wanted these assumptions brought out clearly. At the same time I wanted the students to be active participants in the course; but, if they hadn't taken calculus, how could they understand, say, curvature without taking a derivative?

I wanted to show that "what one knows" is often "what one assumes," and I wanted these assumptions brought out clearly.

It became clear to me that I did not need "exercises" from a text. What I needed were student projects which would explain, in a hands-on way, the concepts of the course. But what projects? I had to invent.

My first invention, and maybe my best, was the following.

> Go to the Cornell Arts Quad. Measure, by any method you want, the height of McGraw Tower, the tallest building on the Quad. Write up your solution on one sheet of paper as if it were a lab report. Use any diagrams you think appropriate to help explain your work.

Students had a number of questions. Can we work together? (Yes, but separate writeups.) Do we have to be neat? (I need to be able to read your work.)

When the project was done, we had a discussion. We classified methods of measurement by type: Pythagorean; similar triangles; estimation; "direct" measurement; nonmathematical models. Then we discussed reasons for possible errors in each method. At this point someone usually began to ask what the "real answer" might be. I would often counter the question with one of my own: "Isn't it yours?"

We also discussed the writeups; in particular, the use of prose versus diagrams, and what meaning that might have for the reader and the writer. For instance, I would lead to questions like: To what extent did the "picture" do a better job of explaining the text than the writing? In what way was the diagram the real report?

We also discussed the "concept" used in the measurement. Did the students "see" a method, then do it; or did they let a method find them? Did their original method fail? How then did they adapt?

Then we went deeper, to the questions behind the questions. Why did you use Pythagoras? Does that theorem really work? Where might it fail? And the same with similar triangles.

Similarly, with respect to conceptual questions: What constitutes research? What does it mean to say you have a method?

Many of the above questions were too advanced for the students at this point, so we went back to more projects. We analyzed elliptic geometry, the geometry of the sphere. We found out what's "straight" on a sphere, and talked about geodesics. We asked about Euclid's original definition of a "straight" line ("that which is even with all points of itself"), and what he could have possibly meant by that. After talking some about triangles, we came back to Pythagoras' theorem, only to find that it fails on a sphere. Then I asked again, "What assumptions did you make about geometry when you measured the height of McGraw Tower?"

Another topic I covered was that of surface. To get the students used to the difference between being "on" something and "in" something, I assigned another project:

> You are a bug who lives in a two-dimensional cylindrical space. First, what do you see? Second, how might you possibly discover that your world is cylindrical, rather than planar?

The above project led to such questions as: How do we know the Earth is round? What possible shapes could the universe have? And how could we discover those shapes, since we live *in*, not on, that three-dimensional universe?

So that the course didn't become "airy generality," I also had a different kind of topic:

> Go to a place you like. Describe it. Now think about some of the geometric words you have used in your description. To what extent do these words have to do with your liking of this place?

I tried not to be too specific about the words students may use in a geometric context; I was also interested in finding out which words they consider to be "geometric."

Assignments came in, and we catalogued words. We concentrated on a discussion of the meanings of some of the more geometric ones we were using in our descriptions, and how those words may have to do with either some aesthetic sensibility, or with a sense of well-being. Then I asked the students to go to the campus art museum to see whether they could use some of the same geometric concepts to describe the art.

Ultimately, we were trying to find a vocabulary from geometry which would explain some part of the aesthetic sensibility.

Now that you have some idea of the projects I assign in my math class, let me change gears slightly, introducing one more topic before I return full circle to the idea of final projects. Last summer at Cornell I taught a writing seminar in the John S. Knight Writing Program. There are various traditional forms for papers given in writing courses, and while specialists argue as to how many such "types" of papers there are, I feel safest using some of these terms: Description, Definition, Narration, Argument, Comparison and Contrast, and the Research Paper, which can combine all of the above.

My first project was the same as in the Exxon course; we measured McGraw Tower. I then had the students read or synopsize their papers, so as to get used to reading aloud in class.

Then we wrote the same paper again, but this time as a letter home (a narrative). This allowed us to think about audience. We then discussed the difference between the simple description given in paper one and the narration in the second.

Project three was an argument paper:

> Some say the Earth is round. Give a proof of this.

We discussed the definition of "round;" what constitutes an "argument" or proof; what is the difference between a hypothesis, an axiom, and a theorem; what a "thesis" is, and so forth.

I am sometimes asked how my "math in writing" class differed from this class. I think the quickest way to describe the difference is to say the following:

> In the mathematics class, I am primarily interested in the mathematics, and use the writing to help explain the geometric concepts; whereas, in the writing class, I concentrate on writing as the topic, with mathematics as the content.

An example from the classroom may help to illustrate this.

One day on my way to my writing class I saw that an art instructor had placed a number of geometric forms on the Quad so that his students could sketch them. I asked the instructor if I could send my class out to write about what they saw while his group sketched. After twenty minutes, my group came back to the classroom where we read our papers. Each student seemed to have taken a different viewpoint; one wrote about the geometric forms themselves, another about the students who were sketching and why, a third about himself watching the students sketching, and yet another about why the instructor may have chosen this assignment.

The class then discussed the use of viewpoint, and how it affected the language used in the essay. For instance, we discussed whether the student who wrote about the geometric forms used more of the passive voice than the others who spoke from a more personal viewpoint. We also asked ourselves for whom we were writing, and how this affected the language we used.

Now I am ready to return to the topic of final projects.

Very early in the semester I announce that the course will end with a project. The form of this announcement is important to me—I do not want students to place the idea of a single major paper at the center of their thought about the course. Rather, I want them to participate in the act of discovery that is at the core of the topic. To this end, at various times during discussions I will suggest that certain questions we are mulling over would lead to "nice projects."

For example: one day we got, not totally unaided, to the question of whether our eyes see in the Euclidean manner. I mentioned that this is a topic on which some research has been done, and that I would be willing to suggest a couple of sources from which to start working.

At about week ten of the semester I have students write two paragraphs:

> "I am thinking of doing my final project on . . ."

> "I may need the following kinds of help . . ."

During the next week I conduct individual meetings of about thirty minutes with each student. This is important to them:

1. The meeting settles and sharpens their theses;

2. The student can get some help from me;

3. The student leaves with the sense that he or she can discuss mathematics with a "professional" without appearing foolish.

The meeting is also important to me:

1. I can discourage or reject any unformed topics, and sharpen any half-formed ones;

2. I can learn about their sources.

In weeks twelve and thirteen I give some final classroom discussions and lectures, usually assigning a reading but no new writing. If there are last-minute questions or project changes I'm available for conferences.

On the last day of classes some students give verbal reports on their projects. I tell them to take ten minutes, but it always takes twenty. The rest of the students finish the reports on another day chosen at mutual convenience. That day we can go as long as two hours.

I then read their written drafts. I have criteria, which apply to earlier papers as well as the final ones. These criteria, in ascending order of importance, are:

1. Grammar;

2. Internal clarity; by which I mean clarity of individual sentences and paragraphs;

3. Clarity of the basic argument or research;

4. Significance of the topic.

I have a standard statement to students about grammar. "No one knows everything about grammar, but everyone should try to learn as much as possible, because when we are trying to communicate good ideas to someone else, we don't want to risk having them rejected because of bad grammar."

I grade papers in a somewhat impressionistic manner. First I mark up grammatical errors I find, simply underlining them. In earlier papers, I have been more specific about problems.

If sentences are unclear, I say so. I offer a suggestion as to how one particular sentence may be changed; then I suggest that the student try to work on a revision of one other problem sentence.

No one knows everything about grammar, but everyone should try to learn as much as possible, because when we are trying to communicate good ideas to someone else, we don't want to risk having them rejected because of bad grammar.

If a basic argument is open to question, I offer some alternative explanation, and ask the student to come discuss the possible reasons with me.

I begin my main comment to the student with some positive statement, leaving my disagreements to later in the text as questions, alternatives, suggestions.

Before I put a grade on anything, I go for a short run to clear my head, then come back and read the paper again. Then I do what I must.

You will want to know what some of my students' recent final projects have been. Some choose artistic questions, for instance, "The Use of Geometry in Cubist Painting." This is a topic which, by the way, came out of the assignment I alluded to earlier of having my writing class go onto the Quad and write about the art students who were sketching. Another such paper is "The Geometry of M. C. Escher's Works." Others discuss the golden ratio. Two years ago a student wrote on the use of space in the Guggenheim Museum and how it affects the way we view works there.

Other papers are on mathematical topics; for instance, on hyperbolic geometry. One particularly good paper from last year gave a new proof, using geometric techniques, that the connected sum of the torus and projective plane is topologically equivalent to the connected sum of three projective planes. Some students build models of regular polyhedra, and discuss various properties of the models. A project from two years ago related a theorem about surface theory to the Euler characteristic.

A type of topic I see often has to do with design. The Guggenheim Museum paper may fall under this category. Another is: What is the impetus, from the geometric viewpoint, behind the "glass house" of Philip Johnson in New Canaan, Connecticut?

Some projects ask about spacetime or relativity or large scale-models of the universe. Conversely, *Flatland* has been the source of a paper.

In conclusion, let me say that I have found the experience of using writing in my geometry class invaluable. I'm not sure how I ever got along without it, and I can only say to you that, although the workload may increase by a linear factor with this kind of teaching, the rewards are sure to go up also—by an exponential factor.

Since I'm a teacher, I can't resist suggesting a homework problem.

In the words of Wallace Stevens, there is a poem called "The Idea of Order at Key West." Consider the following stanza:

> Ramon Fernandez, tell me, if you know,
> Why, when the singing ended and we turned
> Toward the town, tell why the glassy lights,
> The lights in the fishing boats at anchor there,
> As the night descended, tilting in the air,
> Mastered the night and portioned out the sea,
> Fixing emblazoned zones and fiery poles,
> Arranging, deepening, enchanting night.

Your assignment is to find some geometric words in this stanza, and explain what picture they might conjure up in your mind. Then discuss what your picture might have to do with explaining the poem's final stanza, and with the topic of teaching geometry:

> Oh! Blessed rage for order, pale Ramon,
> The maker's rage to order words of the sea,
> Words of the fragrant portals, dimly-starred,
> And of ourselves and of our origins,
> In ghostlier demarkations, keener sounds.

The above stanzas of the poem are strongly concerned with both geometry and the creative process, and speak directly to the core of my course.

REFERENCE

Wallace Stevens, *The Collected Poems,* Vintage, New York, 1982.

A Writing-Intensive Mathematics Course at Western Michigan University

Arthur T. White
Western Michigan University, Kalamazoo, Michigan

Introduction

When I arrived at Western Michigan University twenty years ago, all undergraduate students were required to pass, as a condition for graduation, a college-level writing course. For most students, this was the course English 105, "Thought and Writing." Although teaching exclusively in the Mathematics Department at that time, I had an intimate connection with English 105: my wife taught eighteen sections of the course, over a three-year period. The stated purpose of the course was, and is, for the students "to develop their sense of language as a means of shaping and ordering their experience and ideas, and to develop imagination, thought, organization, and clarity in their written work."

It seemed to us that the course was, at least partially, successful in achieving this purpose. But necessarily the writing experiences it provided were general in nature, rather than being tied to a specific discipline. There was a feeling among the faculty around campus and, I think, particularly in the English Department, that writing experiences would be more relevant for students if tied to course content in their mainstream courses. English 105 continued to be offered, but not as a requirement. The idea was that we were all supposed to integrate meaningful writing experiences into our regular courses, so that the students would be writing in context. The maintenance and enhancement of writing facility for the students was a responsibility to be shared by all faculty, not just those in the English Department. But we were given no guidelines as to how to meet this responsibility.

The idea was that we were all supposed to integrate meaningful writing experiences into our regular courses, so that the students would be writing in context.

For most of the courses I taught in the Mathematics Department at that time, I did not see how to incorporate writing experiences effectively. But in one course, Mathematics 190, "A Survey of Mathematical Ideas," which I first taught for the Honors College in 1973, I assigned one short (3–5 pages) typed paper and one longer (6–10 pages) typed paper on topics of interest to the student and deemed, by myself, to be relevant to the course. Popular topics included the history of a particular mathematical concept, biography of a famous mathematician, math and music, math and art, math and poetry, and the like. Shorter writing exercises were woven into the course. I have taught this course five more times since (and a sixth time for math majors, at a higher level, as Mathematics 490), each time as "writing-intensive." But I think my involvement in this way was fairly unique, certainly in the Mathematics Department.

Beginning in 1982, I have been venturing periodically into the English Department, where they kindly allow me to teach English 107, "Good Books." It is natural to teach this course as "writing-intensive"; we have numerous quizzes (short answer) and study guides, two essays written in class, and two papers (3–5 pages) written—and typed—outside of class. I enjoyed teaching this course again—for the fifth time—in Fall Semester, 1989.

I believe that both "writing to learn" and "learning to write" are important. Thus I grade my writing assignments both for content and for the mechanics of presentation. Specifically, I assign a grade, fairly subjectively, based upon content; then I reduce this grade, fairly objectively, based upon the number of mechanical problems. Some errors—such as sentence fragments and lack of subject/verb agreement—count more heavily than others.

The idea of widely shared responsibility for maintaining and enhancing writing skills through context-relevant experiences certainly has merit, but I cannot say that it was, initially, a resounding success at Western Michigan University. Now I believe that we are proceeding in a sensible manner. Our present writing requirement, for undergraduate students, has two components. First, they must pass a college-level writing course. (This part of the requirement was reimplemented in 1982.) The course is typically our old friend English 105. But alternatives are available, in the College of Business, in the College of Engineering and Applied Sciences, and—in the College of Arts and Sciences—in the Departments of History, Philosophy, and Religion. Weak writers must pass a remedial course first; strong writers can be exempted by examination. Second, they must satisfy the "Baccalaureate-Level Writing Requirement." The preferred way to do this is to pass a course in the major discipline (or in a related alternative) which has been officially designated as "writing-intensive." This designation means that the course will integrate several writing tasks into the term's work and that their evaluation will comprise a significant portion of the course grade. Criteria for evaluation include the ability to: (1) demonstrate maturity of thought, including analysis, synthesis, and evaluation;   (2) sustain the development of a point or idea over at least 500 words;   (3) use effectively organized paragraphs and transitional devices;   (4) use capitalization and punctuation conventionally; and   (5) use regularly (if not faultlessly) standard grammar, syntax, spelling, sentence structure, and agreement between subjects and verbs, pronouns and antecedents.

In the Mathematics Department we have designated two courses as "writing-intensive." One of these, Math 402 ("Mathematical Modeling"), is a requirement for the Applied Mathematics Major Option. The organization of the course around case studies makes it nicely suited for relevant writing experiences. The other writing-intensive mathematics course is 314 ("Mathematical Proofs"); it is required in all three of our other major options (General, Secondary Teaching, and Statistics). I have been teaching Math 314 regularly since its inception in 1980, and I have always taught it as writing-intensive—even before being required to do so—because I have thought it a good (although perhaps eccentric) idea.

In the summer of 1983, by participating in a Writing Across the Curriculum Institute organized by Lynne McCauley, the Director of our Intellectual Skills Development Program, I learned that my "good" idea was, serendipitously, actually a respectable one. Since then my firm impression that writing *does* have a place in the mathematics curriculum has been further reinforced by numerous on-campus activities conducted by Lynne McCauley, by the great interest shown in Andrew Sterrett's special session on "Writing as Part of the Mathematics Curriculum" at the January, 1988 meeting of the Mathematical Association of America in Atlanta, by the response to my presentation at the Michigan Section meeting at Eastern Michigan University in May, 1988, by reading books such as William Zinsser's *Writing to Learn* [5], and—most of all—by my experiences in teaching the Math 314 class.

Sharing these experiences with anyone willing to read this essay is my principal reason for writing it.

Mathematics 314

Here is the catalog description for Math 314, Mathematical Proofs (3 hrs.):

> The prime objective of this course is to involve the students in the writing and presenting of mathematical proofs. The topics in this course will include logic, types of proof, sets, functions, relations, mathematical induction, proofs in an algebraic setting such as divisibility properties of the integers, proofs in an analytic setting such as limits and continuity of functions of one variable. Additional topics may include elementary cardinal number theory, paradoxes, and simple geometric axiom systems.

(As a matter of fact, I wrote that myself.) The prerequisites are two semesters of calculus and some linear algebra. This is not so much for specific content, but in the (sometimes vain) hope that the student will have acquired a modicum of that elusive entity, mathematical maturity. The course is typically taught five times per academic year: twice each in the fall and winter semesters and once in the spring session. The average enrollment is about twenty students. I try to teach two of these sections each year, because I think it is an important course, the course in which I believe I can have the most impact, and the one I most enjoy teaching.

The other instructors for the course have differing means of achieving its objectives. Here is what I do.

I use a modified Moore method of instruction, and I teach the course as writing-intensive. Thus my students are astounded (dismayed?) when they learn that the only text for the course is *Write Right!* by Jan Venolia [3]. (This is a beautifully concise and usable review of grammar, punctuation, and style.) I get frantic calls from students prior to the term, inquiring if they really have purchased the proper text for the course.

Yes, they have. I am trying to inculcate, in each student, a facility for and pride in discovery, creativity, and self-reliance—in the context of mathematical proof. I believe that availability of and reliance upon a mathematical text would defeat these purposes.

I want to explain the modification of the Moore method which I use in this course, and how I think it is ideally suited for a writing-intensive course. First, I should describe the Moore method itself, briefly. (For considerably more detail, see [4].) R. L. Moore taught, at the University of Texas, with great success, courses in calculus, foundations (analysis, geometry, mathematics in general), and topology. He would give his students definitions (often without examples) and statements based on the definitions, and then turn them loose to prove the statements as theorems. Proofs would be presented by the students, orally, and then critiqued by the class. No source was allowed other than each student's own mind. Sometimes the students were expected to introduce definitions and to conjecture statements on their own. Sometimes they were given statements which were, in fact, open research problems. Moore felt "that student is taught the best who is told the least." His method aims to (1) get the student involved, (2) lead the student into personal discoveries, and (3) let each student proceed at his or her own pace. Imagine the thrill of recreating a significant branch of mathematics from your own internal resources! The system seems ideal, for a small group of highly motivated and perhaps even gifted students.

Not being completely confident that I have exactly that combination in my sections of Math 314, I have decided to modify the method. My course is organized around a collection of roughly one hundred mathematical statements, some true and some false. (I include some sample statements, in the Appendix.) For each new topic, I present definitions, examples, and sample statements (not among the one hundred, but each with suitable proof or disproof). Then I assign relevant statements for student consideration. Some of these are to be resolved (the student must decide whether the statement is true or false, and then provide proof or disproof; in some cases an invalid proof of a false statement is given, and the student's task is to find the error in the proof) in writing, and for these no source other than the student's own intellect is allowed—except that I will answer questions and give hints in my office or by phone at home, to facilitate continual progress. Others are to be presented before the class and then critiqued by the class. For these "board" problems, any source is allowed, as long as it is properly acknowledged. Thus the students can do library research, collaborate with each other, or ask me for help (although I try to impress upon them that most of the statements are sufficiently elementary that they can be much more readily resolved by independent thinking than by exhaustive literature search). Thus I try to provide an environment in which some fall back to established resources is allowed, while leading the students gradually into the joys and triumphs of independent discovery. Since the course is primarily process-based rather than content-based, I feel no compunction to speed through a preconceived number of statements, nor to use the same statements term after term. Thus we often digress, to discuss not only the nuances of a particular proof, but also possible generalizations and corollaries of the statement, possible improvements or embellishments of the proof or of its presentation, and possible alternate proofs of the original statement. Each student's final grade for the course has four components: (1) written problems (30%), (2) board problems (30%), (3) final examination (20%), and (4) the writing-intensive aspect (20%). It is this last component which I now want to address specifically.

The written problems are graded for mathematical content only, although I give informal feedback regarding the mechanics of writing by indicating errors (or possible improvements) of spelling, punctuation, diction, and the like. Near the end of the term, I have the students read Uri Leron's article "Structuring Mathematical Proofs" [1]. In this article, Leron contrasts two main styles of proof, the linear proof and the structured proof. I ask each student to describe these two proof styles, giving one good illustration of each from the course, and then saying which method they prefer and why—in 2–4 typed pages. The papers are graded for content and mechanics of writing, and this accounts for one-quarter of the 20% for the writing portion of the course. This assignment is a good example, I believe, of the "writing to learn" mechanism. In writing about the two styles of proof, and especially in carefully writing out a specific proof in each style, most students really learn about these styles and, in the process, teach themselves which is better suited to their own needs. (Many decide that each style has its proper place.) Of course, I try to assure that they "learn to write" as well, by carefully reading their papers for mechanics of writing also.

The rest of the writing-intensive aspect of the course (three-quarters of the 20%) is the part that I am most excited about. Each student keeps a journal, with one entry (hopefully, a substantial, cohesively developed paragraph) for each of the one hundred or so statements they have been wrestling with. They are encouraged to write this entry immediately after the resolution of the statement has been completely discussed in class.

Initially, I was focused on "learning to write," and so I encouraged the students to give, in their paragraphs, their subjective response to their experience with the statement in question. If they resolved it correctly, what was the source of their inspiration? If not, what was the source of their blockage and frustration? In either case, did the statement seem important to them? Did they learn something from struggling with it? Should it be kept in the course, or discarded?

I was amazed to find that many students, despite these encouragements to write an "easy" paragraph—since there were no right or wrong answers to my questions—insisted on writing paragraphs that paraphrased the given statement (and/or its proof), that broke it down into components they could intuit and feel comfortable with. These students were determined to "write to learn," whether I wanted them to or not!

This has helped to convince me that "writing to learn" *really is a good and workable idea*. Students do it naturally, and they profit from it.

These students were determined to "write to learn," whether I wanted them to or not!

Now I let them write about almost anything they want in their paragraphs (it should have some relevance to the statement at hand): subjective responses, paraphrases of the statement and/or its proof, conjectures of corollaries or generalizations, discussions of alternate proofs, questions about the mathematics that trouble them, gripes about the course, and even (shudder!) gripes about me.

This opens a marvelous two-way channel of communication between the instructor and each student. I collect the journals (usually) three times a term, read them carefully, and return them promptly. I try to answer their questions (mathematical and otherwise) and address their concerns, in writing, in the journal. When more extensive discussion is indicated, I suggest an office visit. Hopefully my students get to know me better by this means. I *know* that I learn much more about each of them by their journals than I would otherwise. Formal student evaluations are almost redundant for me in this course; most of the students have already bravely told me what they do and don't like, while I still have time to make adjustments *during* the term in progress. Every time I teach this course, I learn many valuable things from my students. Most of my learning occurs through the journals. I hope—and believe—that most of the students profit from the journals as well. A word of caution: this is a very time-consuming activity (for me, about three weekends each term). But, of course, most worthwhile enterprises have an associated cost.

Student Paragraphs

Many of my students are surprisingly good writers, and most of them have very creative insights that they are willing (in some cases, eager) to express. Here are some sample paragraphs, all from a semester recently completed (Fall, 1988). The five students represented were, in my opinion, all from among the best writers in the course, as evidenced by their clarity, originality, and depth of thought and by their care, precision, and organization in writing. Perhaps not coincidentally, they were among the best students in the class mathematically as well. I thank them all for allowing me to use their paragraphs here. In some cases I am quoting only a portion of a paragraph, but I am not editing in any other way. (The hardest part about writing this essay was to decide which of the many splendid paragraphs by these—and the other—students to *omit*.)

We began the course with truth tables, both to establish logical principles which could be used throughout the course (and beyond) and also to break the ice with proofs that the students could easily produce and soon found to be routine. After several truth tables, Naeem Murr wrote:

Alas! Yet another truth table confronts me. Let me quote H.L. Mencken: "The smallest atom of truth represents some man's bitter toil and agony; for every ponderable chunk of it there is a brave truth-seeker's grave upon some lonely ash-dump and a soul roasting in hell." Obviously Mr. Mencken had also had experience of truth tables. Of course I am being melodramatic again; in fact I managed to run through this truth table in just a few moments.

We discussed the method of indirect proof, including indirect proof by contradiction. Compare the following two paragraphs, by Aaron Mead and Charleen Smith, respectively.

This problem in itself was not very interesting. But I may make an interesting application of it. A couple of years ago, my father asked my sister (aged 7 at the time) whether there could be a biggest number. He expected to hear an argument similar to the proof of W13—that there could be no biggest number because you could always add one to it. Instead, she said that there could be no biggest number because if there was, it would have to be the number of days that heaven lasted—an impossibility, since heaven lasts forever. This is an interesting case of proof by contradiction if I ever heard one, and I think the next time I go home I will ask her if there can be a smallest number. Now that she is nine, I wonder if I will get a mathematical or a theological answer?

If I had to choose a proof form which I like best, it would be the form of the contradiction. To have the freedom just to assume that something which is wrong is right, and thus to find that that which is right is then wrong, and in so doing to prove beyond a shadow of a doubt that everything which is wrong is rightly wrong, is an interesting way of viewing the world. I'm still learning to accept a form of thought where things are either/or. Doesn't someone with a highly trained mathematical mind have at least some trouble functioning in the real world?

I introduced the students to various heuristics that might be used in determining the validity of statements. Nancy Holmes tried three analogies to see if union might distribute over intersection, and then Naeem questioned the process of analogy in general, in these respective paragraphs:

This was my first and most extensive experience with heuristics. It was interesting that only two of the three analogies that I attempted showed that the statement should be true. The Venn diagram and the truth table showed truth, while the analogy with algebra suggested fallacy. In the end, the proof of the statement held, and the truth table turned out to be the best analogy because, after all, we used the logical statement in the proof.

I must admit that analogy perplexes me to a certain extent. Why does it work in some cases and not in others? What is the real basis for an analogy? Is there any justification except that it seems to work most of the time?

Following Richard Guy, I defined a "proof" as "a group of people, a collection of symbols, and a measure of conviction." We always had the first two entities well in evidence, but the third seemed to be variable. Many students wrote about this crucial aspect of proof; here is what Nancy said, in the context of her first board presentation:

This problem marked my first board presentation as well as our first alternate proof, and I think both went over well on the whole. I presumed that I would not be nervous because usually I take public speaking in stride, but when I went up to the board I simply felt tongue-tied. I tried to give some insight and explanation as I wrote my proof on the board because I felt that previous presentations lacked in this respect. Apparently, I was at least partially successful because Dr. White commented about getting insight into the presenter's way of thinking. Next, the class compiled an alternate proof which, personally, I preferred to my own because it was more sophisticated. Altogether, my presentation went fine, and hopefully, my next one will be better.

Later, she addressed this issue again:

This presentation created some controversy over how much detail is necessary to produce a convincing proof. Several people did not want to include the lemma from logic in the set theory proof. Dr. White, however, seemed to think that this step was indispensable, although he simply stated that we leave it in without being absolutely forceful. The conflict was never resolved completely and still represents a *gray area* in our proofs.

I constantly tried to get the students to think about alternate proofs, so as to increase their arsenal of possible approaches. Nancy mentioned this briefly above; Gretchen Zang wrote as follows, about the statement that the product of three consecutive positive integers was always divisible by three:

Three proofs of this statement were presented in class. One was an induction proof, one used partitions and one was a corollary of problem "62." I found the "corollary of '62'" proof. We discussed the pros and cons of each proof and then voted on what we thought should be the "book" proof. I think the induction proof should be the "book" proof because the partition proof requires understanding of partitions and that's not something most people have. The corollary proof requires one to remember problem "62" and as I sit here writing this, less than one week after doing this problem, I don't remember problem "62." I don't think a lot of people would. That leaves the induction proof for the "book."

I think of mathematics as an art form, with aesthetic values to be appreciated; Naeem caught the spirit of this, with two of his paragraphs:

I would be interested to know how a formulation such as $\equiv$ (modulo n) is arrived at. Is it the result of a flash of insight, or is it something more deliberate and mechanical? As a child, I was told about the academic Berlin wall which exists between art and science. Rigorous analysis and creativity are oil and water, they do not mix. I wonder if it is too late for me to consider the extent of this fallacy.

Essentially, this problem was exactly the same as the previous one. Mr. White described the form of an induction proof as parallel to that of a Shakespearian sonnet. The pivot of the proof is the small, but fundamentally significant, word 'but.' Is mathematics poetry? Or is poetry mathematics? Or are these questions the two halves of a dialectic, the synthesis of which few have experienced?

For several of our problems, I presented clearly false statements, with not so clearly fallacious proofs. The idea was to accustom the students to spotting erroneous steps in others' work, so that they would be able to check their own proofs effectively. Aaron had a lovely analogy about his heuristics in debugging an algebraic proof that any two integers are equal, and Charleen made an excellent suggestion that I incorporated into my teaching the following semester.

I carefully went through each step in this false proof several times and could not find the error. So I ran the obvious test: I plugged in different values for x and y and watched to see which line was the first to be false. This is similar to the test I perform on the parts of my clarinet every so often to test for leaky pads. I plug the bottom of each tube section, cover all the holes, and blow air down into it from the top. If there is a leak somewhere, I will sense it immediately and can pinpoint it.

I found the incorrect proofs very difficult, and spent too much time on them and not enough time on the proofs we had to write. On the one hand it was fun to puzzle through them, but after several hours and still no idea what the mistake was, it somehow wasn't fun anymore. At one point I felt that if you wanted $2 = 1$ so badly, I didn't care anymore and it was all right with me. I don't think that that is the attitude that you were hoping to nurture with these exercises. The more difficult ones would have been more fun doing in small groups in class, as we worked out the Towers of Hanoi problem. They would have been less time-consuming and more instructive than trying to figure them out on my own.

We discussed the creative process whenever possible, and the theory that the subconscious is engaged on really tough problems, following periods of intense conscious concentration. Aaron wrote about an exemplary instance of this.

I don't remember ever having had the experience I had with problem 70 with any other problem. I worked on this problem for hours, and finally went to bed at 2:00 a.m. As I was trying to forget the whole thing and go to sleep, the solution suddenly dawned on me. Luckily, I was able to remember it the next day.

Gretchen wrote this perceptive and introspective paragraph near the end of the term:

This is another proof I doubt I will ever need again. I suppose the exercise of thinking about it is beneficial. Personally, I think most of the higher math is outdated and will only be for those who have time on their hands. Computers have already taken over most of math and will continue to do so more and more. I'm not saying that computers are the answer to all of math's problems. I am a practical person and I like math that is useful. Most real world problems that previously needed math for a solution now can be simulated on a computer and solved that way. There is the added benefit that computers can estimate solutions to equations that mathematicians can't or that mathematicians don't have the time to do. Math does make you think, though, and can be a very engrossing pastime. I also think that a good understanding of high school math is needed for a well-rounded person and that is why I want to be a math teacher. I have learned from this class how important it is for a teacher to have a sense of humor and to constantly ask the class questions. This is one of the first classes I've taken that I've kept on track during class time. Usually I just blindly copy down notes and then try to figure out what went on afterwards. Sometimes that works and sometimes it doesn't. Thinking during class is a new concept for me. I like it. I'm going to try to involve the class as much as possible and try to let my sense of humor show through when I become a teacher.

Charleen added an extra paragraph at the end of her journal, that I was very pleased to read. A few students were considerably less enamored with the journal idea, but I believe that Charleen expressed, perhaps more eloquently than most, the predominant opinion in the class.

> Working on this journal has been a real pleasure, and one of the more interesting aspects of a very interesting and challenging course. I had never done anything like it before in a course, and have been very surprised at how much it has forced me to focus on the material, instead of just completing a problem and forgetting it afterwards. Here I have had to rethink problems, remember my approach, and how I have worked my way through them, or not, and the class discussion about them. They must be very revealing for each of us students, and I sometimes worried about that, but it's difficult to write this amount without going deeper and deeper into yourself just to find more material. But they must also, taken together, be a very good indicator of how the class is working. And, as you yourself said, you continue to do fine-tuning of the instructing on the basis of the journals. This has been a good class!

Lynn Arthur Steen [2] wrote as follows:

> Effective teaching for today's students requires a more diverse repertoire of approaches, including in addition to lectures, homework, and examinations, new opportunities for group work, for extensive writing, for oral practice, for exploration and experimenting, for modelling projects, and for computer activities. Students need to experience mathematics as they learn it, and not simply study it in preparation for exams.

My experience indicates a strong agreement with Professor Steen's advice, especially as it pertains to an abstract course such as "Mathematical Proofs." I have not attempted to incorporate modeling projects into the course. (Our other writing-intensive course— "Mathematical Modeling"—is designed specifically for this.) Many of my students voluntarily involve the computer in the experimentation stage of their attacks upon the statements I give them. The other aspects of Professor Steen's remarks reinforce what I have been trying to accomplish in Math 314. In particular, the writing component of such a course, although challenging for both student and instructor, can be both valuable and enjoyable for all concerned.

Appendix

(Some statements used in Mathematics 314, at Western Michigan University.)

1. $\sum_{i=1}^{n}(2i-1) = $ _____ (complete, with proof).

2. $n^2 - n + 41$ is prime, for all $n \in N$.

3. If x is a 2-digit number, and y results from x by interchanging the two digits, then $x - y$ is divisible by 9.

4. Implication is associative; that is, $[(p \Rightarrow q) \Rightarrow r] \Leftrightarrow [p \Rightarrow (q \Rightarrow r)]$.

5. $\overline{A \cup B} = \overline{A} \cap \overline{B}$.

6. The natural numbers are in one-to-one correspondence with the positive reals.

7. A 2×12 chessboard can be evenly covered by 2×1 (or 1×2) dominoes in exactly 233 different ways.

8. There is a smallest positive rational number.

9. The three defining properties for an equivalence relation are logically independent.

10. An equivalence relation on a set provides a partition of that set.

11. The digits $0, 1, 2, \ldots, 9$ can be ordered, without repetition or omission, so as to form a prime number.

12. $(d^n/dx^n)(1/x) = (-1)^n n! x^{-(n+1)}$.

13. San Francisco is an easy walk from New York. (Find the error in the "proof.")

14. The Binomial Theorem.

15. The Laplace transform of t^n is $n!/s^{n+1}$.

16. The Euclidean plane can be tessellated by regular n-gons if and only if $n = $ _____ (complete, with proof).

17. All polynomial functions are continuous everywhere.

18. There is a projective plane of order 7.

19. The union of two convex sets is convex.

20. Arbitrary unions of open sets are open.

References

1. Uri Leron, "Structuring Mathematical Proofs," *American Mathematical Monthly* 90 (1983): 174–185.

2. Lynn Arthur Steen, "Mathematics For a New Century," *Notices of the American Mathematical Society* 36 (1989): 133–138.

3. Jan Venolia, *Write Right!*, revised edition, Ten Speed Press, Berkeley, California, 1988.

4. Lucille S. Whyburn, "Student Oriented Teaching—The Moore Method," *American Mathematical Monthly* 77 (1970): 351–359.

5. William Zinsser, *Writing to Learn*, Harper & Row, New York, 1988.

Writing, Teaching, and Learning in Mathematics: One Set of Experiences

Richard J. Maher
Loyola University of Chicago, Chicago, Illinois

The Department of Mathematical Sciences of Loyola University of Chicago offers a number of undergraduate programs in mathematics, statistics, and computer science. The department is a unit in the university's College of Arts and Sciences, which has an extensive "core curriculum" in the humanities, social sciences, and natural sciences that is required of all undergraduates in the college. Unless exempted, all freshmen also take a two-semester writing sequence in the English Department. Ongoing discussions over the past few years have led to the introduction of "writing intensive" courses, both within majors and in various core courses. While our department is not at this time required to offer such courses, which include, in particular, a large number of writing assignments, there naturally has been a good deal of discussion about them. This discussion in turn has raised questions that we all must address concerning the relationship between writing, teaching, and learning in the mathematical sciences.

My department's first significant encounter with the writing-teaching-learning relationship did not arise from the circumstances outlined above. In the mid 1970's, it became apparent that too many students in our mathematics and mathematics/computer science degree programs were having difficulty making the transition from the two year calculus sequence to the introductory sequences in analysis and algebra. Their main difficulty involved the concept of proof. As a result, the department instituted a seminar course dealing with the construction and analysis of proofs. This course carries two hours credit, is limited to ten students per section, and usually is taken by sophomores. The students prepare material for class presentations in conjunction with the faculty member supervising the course; they have access to other faculty if needed. While the original purpose of this collaboration was to insure the correctness of the material to be presented, it now also involves the form it takes. Grading normally is based upon the presentations and upon required written assignments, not presented in class, that are derived from class material. (A more detailed description of the seminar is given in R. Reisel, "How to Construct and Analyze Proofs—A Seminar Course," *American Mathematical Monthly*, 89 (1982), 490–492).

The course has in fact eased the transition to analysis and algebra.

The results of this course have been quite gratifying. The quality of mathematics prepared improves during the semester, as does the quality of the writing. There seems to be general agreement that well-written mathematics is also well-learned mathematics and that the students feel a certain sense of accomplishment in what they have done. Furthermore, the course has in fact eased the transition to analysis and algebra. Finally, while it was not planned that way, this course will serve as a writing intensive course, if needed, for majors in our programs in mathematics, mathematics/computer science, and mathematics/statistics.

A number of our majors are students whose native language is not English or whose high school preparation is somewhat less than ideal. Some concern was expressed initially over whether these students would be able to handle a course that involves a good deal of mathematical writing. These concerns have not proved well-founded. Our seminar students are sophomores who have taken two semesters of writing in their freshman year. The students mentioned above normally are assigned to special sections of these writing courses, where extra resources are available to them. Since most of our majors take Public Speaking as a core curriculum course, they rarely encounter problems when talking in front of a group. As a result, performance in the seminar course seems to be a function only of the students' own abilities.

One final point should be made concerning this course. Faculty being assigned this seminar are not being given a reduced teaching load. While there is not much preparation needed, there is a good deal of student contact. I personally have found that this course requires as much time as any undergraduate course that I have taught.

Our seminar course has involved the whole department in the writing-teaching-learning relationship in one particular mathematics course. However, a number of our faculty have used writing in a less extensive fashion as a tool for learning in various upper division courses. In particular, the author has used writing in real analysis, linear algebra, statistics, and operations research courses. The real analysis course is a proof oriented course that by its nature requires writing. The linear algebra section taught by the author is for computer science majors and has a different orientation than the one offered for mathematics majors. Thus the comments below involve only the courses in statistics and operations research.

The statistics course is a one semester course with a one year calculus prerequisite. This course is required of all majors in our department; it also enrolls a few biology, physics, and social science majors. It serves as a prerequisite to a number of courses, including a one year sequence in probability and statistics. The first half of the course provides the background in probability, combinatorics, and random variables needed to devote the remaining time to statistical inference. The students know from the beginning of the term that both their homework assignments and their examinations may contain essay-type questions. They also know that the answers to these questions will require not only an understanding of the mathematics involved but also the ability to present this mathematics in a reasonable, written fashion. Finally, they know that the answers to these questions will be graded on both their form and their content.

There are many course areas amenable to questions of this type. Early in the course, for example, the students might be asked to describe the difference between mutually exclusive and independent events or to distinguish between an event that is certain and an event which has probability one. Later on, questions might be raised concerning the relationship between confidence intervals and probability or about the design or suitability of a particular statistical test. Note however that questions need not involve lengthy discussions in order to incorporate writing. For example, any question or problem that can include an expression such as "Explain your assertion," provides an opportunity to make a judgment both on the mathematics and on the way in which it is written. There also are opportunities to use writing as an aid to learning by means of "projects" given during the semester; some comments on such projects are made below.

The operations research course is a one semester course having calculus, linear algebra, statistics, and programming as prerequisites. It is an elective course open to all our majors and normally is taken by juniors and seniors. The first half of the course is devoted to linear programming, including the simplex method, duality, and sensitivity analysis. The second half is flexible in content and may incorporate student suggestions on what material should be covered. Topics that the author has discussed include queueing theory, network analysis and project planning, game theory, special linear programming problems, decision analysis, and inventory models. As in statistics, there are many topics in operations research that can involve the writing-learning process. Questions about the strengths and weaknesses of various operations research models or about the modeling process itself also can be raised.

An operations research course provides an interesting opportunity for the use of writing in learning mathematics. The history of operations research is a fascinating subject that is accessible from readily available sources. Fifteen years ago it could be the subject of several lectures in a reasonably comprehensive course. Today it no longer is possible to allocate the needed time. This is unfortunate, since, outside of a history of mathematics course or a specialized seminar, students rarely have the opportunity to gain an appreciation of the history and development of a specific discipline in the mathematical sciences. Since a knowledge of the history of operations research can be useful in understanding components of its continuing development, it is entirely reasonable to assign a report on this topic as part of an operations research course. The bibliographies of most operations research books provide readily available sources for the information needed to prepare such a report.

The idea of possibly requiring a written report in an operations research course raises the whole question of using lengthy writing assignments or projects as an aid to learning mathematics in undergraduate major courses. Such activities might be supportive or illustrative in nature (a statistics project where data is collected, described, and analyzed; hypotheses tested; and the whole process then critiqued; with all this done in the order in which the text or instructor presents the topics) or might cover topics not discussed explicitly in class but on which source material is readily available (the history paper in operations research). The author has used such assignments and projects and has drawn a number of conclusions (i.e., has a number of opinions) concerning their use.

a. Such assignments or projects should never be due at the end of a term unless they are a continuing component of a two term course. Students need feedback on what they have done, as well as a chance either to use or to improve upon the material that they have prepared.

b. Such assignments or projects should not be assigned and then forgotten. Periodic reports should be required. These reports should be well organized and well written, with grading based in part upon their content. (Such reports do simplify the student's work in the long term.)

c. Such assignments or projects can be beneficial to the learning process. They also can increase the teacher's work load dramatically. And, they may not always attain the desired learning goal; i.e., they need not always work.

In general, the question of where writing fits into the overall process of learning mathematics is one that must be addressed in great detail. There certainly exists a body of opinion holding that this question can be answered with one word: nowhere. It also seems clear that at present there is no collective body of experience that either confirms or denies the accuracy of such an assertion. Finally, this general question seems to lead to a number of other questions that should be discussed. For example, must distinctions be made among mathematics, statistics, and computer science courses? Is the quality of the writing really that important? Is the level of the course important? Can criteria be developed that will help determine whether writing activity might be profitable in a particular instance? Or, must the approach be through trial and error? How clear is the line between the use of writing to enhance learning and the use of writing to test learning? Perhaps the contents of this volume will offer some direction on all these questions.

It is the author's experience that, in appropriate settings, writing is supportive of the mathematics learning process and that the quality of writing is important. The seminar course mentioned at the beginning of this article provides an example of such a setting. There also are other mathematics courses in which writing seems helpful to the learning process. The realization that they will have to present something mathematical "in writing" motivates the students in these courses to better organize their thoughts from the beginning. This improved organization in turn seems to lead these students to a better understanding of the material they are studying. There may well be substance to the often heard dictum to "Put it in writing." Students who put it in writing in the right places do seem to learn better.

If students develop writing skills at some point during their college education, then it is not unreasonable to expect them to utilize these skills in later courses, no matter what the discipline.

The above discussion does not mean that mathematics courses should become writing courses; nobody is interested in form without substance and we are, after all, talking about mathematics courses. However, if students develop writing skills at some point during their college education, then it is not unreasonable to expect them to utilize these skills in later courses, no matter what the discipline. If, in certain settings, writing mathematics, and writing it well, helps students to learn mathematics, and to learn it well, then writing should be used. The author feels that there are occasions when the use of writing is appropriate.

A number of topics that arise naturally during any general discussion of the writing-teaching-learning relationship in mathematics have not been commented on specifically in this article. These topics seem to fit into three distinct groupings.

1. What happens in lower division major courses? If in fact writing is an effective tool in some courses, when is the earliest it can be used? The freshman year is out in most cases, but what about, for example, the second year of calculus? Is its content amenable to writing? Can writing be effective? Is writing a topic that should be considered by those now discussing the reform of the calculus sequence?

2. What happens in computer science courses? If such courses are offered in a mathematics or mathematical sciences department (as they are at Loyola University of Chicago) and if writing is an effective tool in learning mathematics, then we also should ask about the value of writing in computer science courses. For example, courses in programming languages and systems software might prove quite amenable to the use of writing. Questions concerning technical writing and program documentation also may need to be addressed. Finally, the use of writing in lower division courses must be discussed; discrete mathematics courses for freshmen or sophomores are of particular interest.

3. What happens in nonmajor courses? If such courses are open to all undergraduates, there can be problems with backgrounds. Nevertheless, if writing really is a learning tool for majors in certain cases, can it also be used in selected nonmajor courses? And what about the role of writing, as in more words and less notation, in nonmajor courses? For example, in some sections of our Fundamentals of Statistics course, which has an algebra/geometry prerequisite and is open to all undergraduates, we use a text that deemphasizes the use of formulas and stresses explanation through language as opposed to symbols. As such, this text is a useful counterpoint to methods-oriented texts that the social science students in these sections encounter in their major departments. (Note: The text is Friedman, et.al., *Statistics*, Norton, Boston, 1980; this mention constitutes neither a departmental nor a personal endorsement.) The publisher has indicated that this text is a good and consistent seller, which means that there is a segment of the basic statistics market that finds this book attractive. Are there similar books either for this course or for other nonmajor courses? In general, is there a place for this type of writing and this type of text in nonmajor courses?

This article has discussed the relationship between writing, teaching, and learning in certain courses taken by mathematics majors. It has tried to describe some of the experiences of one department and one of its faculty with this relationship and has noted some questions that need to be addressed. Perhaps it also has provided some information that can be of use in the general study now taking place on the relationship between writing, teaching, and learning in the mathematical sciences.

Technical Writing for Mathematics Projects

J. Douglas Faires
Department of Mathematical and Computer Science
Youngstown State University, Youngstown, Ohio

and

Charles A. Nelson
Department of English
Youngstown State University, Youngstown, Ohio

I. The Common Problem and Its Solution

In *Writing Well for the Technical Professions* [1], Eisenberg explains how "Studies of technical professionals show they typically spend anywhere from a quarter to a third of their workweek using communication skills." Houp and Pearsall in *Reporting Technical Information* [3] confirm the central role communication skills have in the technical professions. Their work cites a study that found engineers spend 24% of their week writing and 31% of their time reading other people's writing. In addition, almost all feel that writing well helped get them promoted and they, in turn, consider the writing ability of subordinates when recommending them for advancement.

Students who see themselves after graduation working merrily away on interesting projects with little regard for their rhetorical responsibilities are simply deluded. How well they can analyze an audience and then make the appropriate adaptations in their communication will help determine their project's success. Besides matching message to the audience, they must be clear about the purpose of their writing and must communicate in a manner free of major as well as petty illiteracies. After all, if they can't negotiate their native language (so the assumption goes), how can they be entrusted with the greater responsibilities associated with more complex and ambitious assignments. How well graduates write and speak affects their professional credibility.

Students who see themselves after graduation working merrily away on interesting projects with little regard for their rhetorical responsibilities are simply deluded. How well they can analyze an audience and then make the appropriate adaptations in their communication will help determine their project's success.

The Youngstown State University Department of Mathematical and Computer Sciences addressed this problem by requiring a senior project of all Bachelor of Science students. Seniors must write a substantial paper in an area of mathematics or computer science and give an oral presentation either on campus or at one of the professional meetings that solicit student papers. This requirement, however, was only partially successful as a professional communication training technique. Project directors are often hesitant to address matters of English usage and organization. Some project directors look on composition theory and instruction as realms outside their areas of expertise. Hence the quality of the student writing and even the format of the papers varies considerably.

Our collaborative effort addresses a problem shared by technically oriented students at many institutions and in many disciplines—the need for students to acquire, in a humane manner, those world-of-work communication skills necessary to facilitate advancement beyond entry-level positions. Our project also addresses a need of our technical communication majors. While few sane English faculty hunger for more papers to read, students majoring in scientific and technical writing and editing—and there has been a proliferation of new professional communication programs in the last few years—need practice editing writing addressed to expert technical audiences. The joint effort between the English Department and the Department of Mathematical and Computer Science gives mathematics and computer science students supervised assistance with their writing projects while simultaneously enhancing the experience of the students working towards the Professional Writing and Editing major in the Department of English.

II. The Method

The easiest way to describe our venture is to outline the new procedures to be followed by students preparing senior project papers. These students, senior mathematics or computer science majors, first enroll in a two-credit, upper-division course offered by the Department of Mathematics and Computer Science, a course entitled Senior Project. As students are deciding on paper topics, securing concept approval, and establishing deadlines with their project directors, they are also being introduced to LaTeX, a set of macros built on TEX primitives.

TEX is not a word-processing program; it is a set of formatting commands that can be embedded in the ASCII text generated by a word-processing program. (We do require students to use a DOS platform, but other versions of TEX support, for example, Apple Macintoshes and Sun workstations.) Portability is a major advantage as it permits students to use the word-processing program of their choice. In addition, TEX is quickly becoming an international standard for the production of technical documents. The quality of TEX documents is unsurpassed, and additional advantages of the TEX system include the ability to represent virtually any technical symbol in use in scientific documents and to print the same document on printers ranging from dot matrix to laser to professional typesetter.

Leslie Lamport's *LaTeX: A Document Preparation System* [5] provides complete documentation for the formatting program. Although this text, together with Donald Knuth's *The TEXbook* [4], is an essential reference for those administering the program, we have found students need only a reduced instruction set. They are provided with examples of TEX format commands placed side-by-side with the output that these commands produce, together with a quick reference list describing just those features of LaTeX needed for their papers. For example, they can consult the manual to quickly discover how to generate bulleted lists, mathematical formulas containing any symbol they might need, matrices, arrays, and tables. An added bonus is that much of the high-quality software needed to implement the system is in the public domain and consequently available at minimal cost, particularly to academic users.

We further simplify the formatting concerns of our students by providing them with a template file for their papers. This file contains comment lines that explain how to enter the text into a format that closely conforms to that required for papers submitted to journals in mathematics and computer science. Following directions provided by comment lines in the file, students need only place such text as headings, subheadings, and body copy in the page mold that is provided, and LaTeX will act as the graphic designer—automatically taking care of such matters as point sizes, white space, hyphenation, and similar visual concerns. Students are free to concentrate on the accuracy, clarity, and organization of their text.

Once students have their topics approved and first drafts ready, they convert their word-processing files into ASCII character files and send them to terminals controlled by the student technical writing consultants. The file is stored until a student technical writing consultant is available. The consultant then makes usage and organizational suggestions, as well as substantive queries, to improve the presentation of the work.

The consultants only embed general comments in the text files to alert authors of the need for certain kinds of revision. For example, while writing consultants do not insert needed transitional expressions, they do call attention to the need for such transitions between sentences or paragraphs and refer authors to the appropriate section of the *Harbrace College Handbook* [2], the latest edition of which is our official style book. The writing consultants are juniors or seniors majoring in professional writing and editing who have completed a course in professional editing. Their work on the project papers is part of the course work required for an upper division technical writing course that is under the supervision of the Program Coordinator in the Department of English.

After papers receive their first edits, their authors retrieve the edited files, make the suggested modifications, and resubmit the file to the student technical writing consultants. This cycle continues until the paper is acceptable to its student consultant editor. At this time the ASCII file containing the TₑX commands is processed. The resulting device independent file enables the writer to view on the computer screen a copy of the paper as it would appear in print. If at this point the project student makes significant changes to the document, it once again undergoes an editorial review. When the project student decides the paper is satisfactory, the file is again sent to the student technical writing consultants with the instruction that the paper be printed.

Project students then take their papers to the appropriate project advisor in the Department of Mathematical and Computer Sciences for approval or suggested modification. Since the paper is editorially correct and in the required format, the only changes at this point concern technical accuracy, development, significance, and insight. After the author makes any modifications required by the project supervisor, the final paper is printed and submitted. Major modifications would, of course, necessitate beginning the editorial review procedure anew.

Thus far we have addressed the writing needs of only one department, but the potential application of the program is much broader. The procedure for submitting senior projects in the Department of Mathematical and Computer Sciences as outlined above could, for example, be expanded to include other disciplines in science, as well as in engineering. In fact, once fully operational, the system could be used by any student or faculty member needing assistance in technical writing.

REFERENCES

1.   A. Eisenberg, *Writing Well for the Technical Professions,* Harper & Row, New York, 1989.

2.   J. Hodges, *Harbrace College Handbook*, tenth edition, Harcourt, Brace, Jovanovich, San Diego, 1986.

3.   K. Houp and T. Pearsall, *Reporting Technical Information,* fifth edition, Macmillan, New York, 1984.

4.   D. Knuth, *The TₑXbook,* Addison-Wesley, Reading, Massachusetts, 1984.

5.   L. Lamport, *A Document Preparation System*: *LaTeX User's Guide,* Addison-Wesley, Reading, Massachusetts, 1986.

ON GRADING

or

"to English or not to English"

But This Is NOT an English Class

André Michelle Lubecke
Lander College, Greenwood, South Carolina

The Motivation

I was so tired of students approaching my introductory statistics course as if it were just "another boring math class," that I decided to do something about it. On the first day of class, I administered a "Classroom Involvement Survey" to see what level of involvement they were accustomed to having in their classes, what level they expected to have in a "math" class, what level they thought would most likely foster their learning, and what level they would most like to have in this particular class. They responded using a 0–6 scale which I developed for this purpose. A score of zero meant that they were "mentally elsewhere." A score of three described them as "actively listening for the purpose of note-taking," and a score of six had them "actively involved in some activity which stimulates thinking, discovery, or learning."

The results were very much what I expected: lots of low scores for what they were accustomed to having and lots of high scores for what they would like to have. On the second day of class, I warned them that if they had lied about the level of involvement they wanted just to impress me, they had better change sections because I intended to keep them involved in lots of "hands-on" experiences.

Most of the classroom activities I planned worked well even though Clyde's Law was always in effect. (Clyde's Law: Everything takes longer than you think it will, even when you take Clyde's Law into account.) My students responded positively to using data which they had collected to discover graphical techniques for displaying data, to learn to calculate probabilities and measures of central tendency and variability, and they liked rolling dice to explore the relationship between probability distributions and sampling distributions, but they were still totally unaware of the richness of the field of applications of what they were learning. What could I do to expose them to real-life statistical problems which were enlightening, interesting, and possibly relevant to their fields of study?

I didn't want them to just *read* something; I wanted them to react to what they read and to spend some time thinking about it.

I decided to browse the library bookshelves to see what kinds of books they held which were written *about* statistics but which were not textbooks. I came across some very interesting titles, and I read a number of chapters in each to see if they were "reader friendly." Many of them were, but how could I use them effectively? I wanted to expose my students to a variety of topics in a short amount of time, and I needed to find a way to do this that would be meaningful for them and manageable for me. I didn't want them to just *read* something; I wanted them to react to what they read and to spend some time thinking about it.

After a great deal of thought, I decided that I would have them read chapters instead of entire books. I identified chapters which could be understood by just about anyone, regardless of statistical background, or those which could be understood by my students because they contained material which was either related to something we had covered in class or a natural extension of it. I also deliberately chose some chapters whose titles contained words which elicited an emotional response, for example, "sports," "discrimination," "nuclear reactor."

But how many should they read? Two chapters seemed too few; someone could conceivably "stumble upon" two chapters which they considered exceptionally boring, and this would not be good. Considering the chapters I had identified as candidates for reading, I decided that one probably could not read four of them without being exposed to at least two different kinds of applications, and I also thought it unlikely that students would choose to read four chapters from the same book, so, four became the magic number.

Much to my surprise and delight, this choice seemed to work well. By the second or third assignment, a number of papers contained comments like: "Everyone should have to read this book," "I never knew you could use statistics in my field," "This really made me think . . . I'll be wary of advertising claims after this," "I wish our textbook were this understandable," "I'm looking forward to reading the next one."

For a few of my students, the readings they selected helped them understand some of the concepts we had covered in the classroom, but for most of them this was not the case. Since my primary objective was to expose them to a subject area in a manner which they would never have discovered on their own, this really didn't matter. I was very satisfied with the results because many of the responses I received showed that they had actually spent time thinking about what they had read. A few "created" applications of quality control procedures which were appropriate for use in job situations with which they were familiar, and one even surprised me with a paper entitled "Statistics on the Rifle Range" in which he described how he was going to use statistics for improving his instruction techniques.

I was impressed! My students were thinking, which is what I wanted, and some of them were having their eyes opened to an entirely new world. All things considered, this was a very worthwhile experience for both myself and my students, and I intend to use this assignment again.

The Assignment

The attached sheet contains a list of selected chapters from four different books which are on reserve in the library. Before the end of the semester, you must read and respond, in writing, to at least four of the selections that are on the list.

This is the procedure that **must** be followed in order to complete each assignment:

1. When you go to the library, make sure that you bring at least one 8 1/2 x 11 piece of paper with you. **As soon as** you have finished reading your selection, begin **writing** in response to it. **Write continuously** for at least five minutes and **do not go back to edit your work.** Don't worry about grammar or spelling errors. Follow your thoughts wherever they may lead. You may find yourself making associations or speculations, drawing conclusions or asking questions. Anything you are thinking is appropriate to write down at this time. The only thing you **may not do** is summarize the article. (Note taking and/or summarizing should be done either during your reading or after this step is completed.) When five minutes have elapsed, you may stop writing.

2. After you have had some time to sort out your thoughts, write a 1 1/2 to 2 page (type-written, double-spaced) response to what you have read. You may re-read your initial response and develop one or two of the ideas you find there, or you may choose to embark upon an entirely different line of thought.

You may not summarize the article!

Summary papers **will not** be accepted. I want to know what you thought about what you read; I already know what it says.

3. Staple together the writings generated in steps 1 and 2 and turn them in to me.

NOTES:

1. You are free to read and write initial responses to any of the selections at any time you choose, but you may not turn in another written response until you have received any previous ones back from me.

2. Due dates for the four assignments are given below. You may turn in an assignment on or before its due date subject to the restriction in Note 1.

Due dates: February 8, March 15, March 29, April 14

OUTSIDE READING LIST

The Statistical Exorcist, Hollander and Proschan

■ Monitoring a Nuclear Reactor, page 27

■ Statistical Control Charts, page 41

Statistics: A Guide to the Unkown, edited by Judith M. Tanur

■ Parking Tickets and Missing Women, page 102

■ How Accountants Save Money by Sampling, page 203

■ Making Things Right, page 229

■ Statistics, Sports and Some Other Things, page 244

■ How to Count Better, page 276

How to Lie with Statistics, Huff

■ The One-Dimensional Picture, page 66

■ Post Hoc Rides Again, page 87

■ How to Statisculate, page 100

■ How to Talk Back to a Statistic, page 122

How to Tell the Liars from the Statisticians, Hooke

■ When the Truth is not the Whole Truth, Ceteris Paribus, Statistics of Discrimination, A Paradox in Discrimination Statistics, pp 19–26

The Rest of the Story

It would not be fair for me to not share with you some of the frustrations I encountered while reading and grading these assignments. I had to find ways of dealing with problems I had never encountered before. With their first assignment,

1. About a dozen students *did not* follow the instructions given, even as clear as they were. Depending on the particular offense, I either required them to redo the entire assignment or to fix what was wrong before I would accept it for evaluation purposes. Some of them had decided that their papers did not need to be type-written, and they had to discover that I meant what I had said. Others had decided to not do the prewriting, and some had submitted papers which were only summaries. These last two groups of students were required to start all over again using a different article. (Word of the severity of the consequences of not following instructions must have spread, because I did not have the same kinds of problems with any of the subsequent assignments.)

2. Of the fifty students in my two classes, all but ten of them were required to seriously revise and/or edit their work *after* my first reading of it before they received credit for it. (On a four-point scale, a piece had to receive a grade of at least a three before it was considered completed.) I had anticipated that this would happen and had allotted ample time between the due dates of the first two assignments to allow for this re-writing to occur.

My students could not seem to understand why I considered the words they used and the manner in which they used them to be important. They seemed to feel that if I could "figure out" what they meant, that that was good enough because this was "not an English class." Once they realized, however, that I expected well-developed ideas which were clearly written without a multitude of grammatical or spelling errors, most of them adjusted their behavior to my expectations and subsequent writing assignments were more frequently accepted without major revisions. (I must admit that I had to work at finding creative ways to respond to their incoherence and/or fuzzy thinking so that revisions did not look like my comments typed into their papers.)

3. A number of students did not turn in what I thought should have been five minutes worth of pre-writing, so I reserved ten minutes of class time one day for a "writing experiment." I gave them a paragraph starter and told them to continue writing until I called time. While they were doing this, I kept track of the time and wandered around the classroom reminding them to keep writing, even if the only thing they could think of to write was "I can't think of anything to write." For most of them, time was not a problem in this situation because I had started them with the phrase "When I first heard that I was going to have to do outside readings in my stat class I . . ." and they had a lot to say about it! Most of their comments were negative and many of them began with words like ". . . I thought you were crazy," but I expected this and tried to keep my spirits up. After all, they had only completed one assignment at this point and many of them did not like having to revise what they considered to be a "completed" assignment.

They seemed to feel that if I could "figure out" what they meant, that that was good enough because this was "not an English class."

By the time they were on their third assignment, most of them had become comfortable with the idea of the pre-writing and I was beginning to receive some very interesting and insightful statements, as well as some positive feedback. They began to realize, from the questions and comments I made on their papers, that I was interested in what they were *thinking*, and that they did not have to try to (or had been told very specifically not to) impress me with "statistical" language, particularly if they were going to use this language improperly.

Along with the chagrin I experienced over their inappropriate use of statistical terms, I was also occasionally genuinely amused by some of their comments. I read about "a pregnant six-month woman," the fact that an "employer was being discriminant," "a nutritioned person," an "average-knowledged person," and a claim that "We need to come to terms where all life forms have an equal chance to succeed in all areas."

I don't know enough about other areas of mathematics to know if these delightful kinds of books are available there as well, but I am very glad I came across the books I did and I am convinced that requiring this type of response, rather than allowing just summary writing, was of great benefit to a number of my students. I watched some of them develop as thinkers and writers, and if I caused even a few of them to think in ways in which they never had before, it was well worth the time and effort.

Let me close by sharing a few of my students' thoughts expressed in their own words:

> "When I first registered to take this class I never thought I would use statistics, but after reading these chapters in these couple of books I now realize how important statistics is in everyday life. I have learned that statistics is used in the fields of medicine, business, sports, and now in law ... I believe that these writing assignments are very worthwhile because they informed not only me, but a lot of my friends in this class about how important statistics really is. It also helped us realize that statistics will play an important role in all of our careers."

> "I am so glad that I have read this article [The One-Dimensional Picture]. It has enlightened me more than I realized an article could. It has been helpful not only where statistics is concerned but also my everyday life has been enhanced. I am glad that I chose this article. I hope I will be less vulnerable to this kind of fraud now. I think everyone should read this article whether they report on it or not. Everyone will learn a valuable lesson from it."

> "I enjoyed this reading [Post Hoc Rides Again]. It made me think about myself and how I handle situations like this. These statistical writings that were assigned to us seemed like a chore to begin with. But now I am enjoying them. It is definitely something I would not have chosen for myself. But I am encouraged to read more. Thank you for opening my eyes to a new area of interest."

You Can and Should Get Your Students to Write in Sentences

Melvin Henriksen
Harvey Mudd College, Claremont, California

For nearly two decades I have insisted that the students in mathematics courses that I teach write in complete English sentences on all written work that I grade. I do this in *every* course; not just in courses for majors in mathematics or courses in which writing is emphasized. While it is rare that I teach courses below the level of the calculus, I do this in such courses as well. I do not go as far as to insist on good paragraph structure or style, but I never read, much less grade work not written in sentences. I point out that symbols are used to abbreviate words or phrases and may be parts of a properly constructed sentence. Thus, "$x = 2$" is a sentence with subject "x," verb "$=$," and object "2" while "$2x = 4 = x = 2$" is not. I insist, also, that any symbols introduced by the writer whose meaning is not clear from the context be defined or quantified properly.

You do not have to teach your students how to write or recognize a simple English sentence unless some of them are foreigners with a serious language deficiency, in which case they should not be in an ordinary college class, and some sort of extraordinary help will be needed. What I am saying does not apply to such students, but does apply to almost all others, independent of their mathematical talents. They can write in sentences, but typically, they will not unless you *really* insist!

On the first day of class I spend some time explaining what the course is all about, what my policy is on homework, quizzes, midterm exams, the final exam, and how I will determine their final grade. I indicate at that time that I am more interested in how they arrive at answers to problems than I am in the answer itself (which often may be found in the back of the textbook), and that I will not read their work unless it is explained *with the aid of complete English sentences*. I stress that I will read what they write and will not try to guess at what they really mean, and that I do not separate form and content. I tell them that I will not grade their English; I just won't read work written in a private code. To reinforce this, I summarize it in a two or three page handout including examples of satisfactory and unsatisfactory solutions of problems which contain explanations of why they are unsatisfactory. One that I have used for a linear algebra course required of all students at Harvey Mudd College appears below.

The initial reaction of most of the students is to ignore both my talk on writing and the handout. They are identified with all the other copy book maxims of education and exhortations given by their teachers about which they need not do anything. Sentences are for courses in English and papers written for courses in history or economics or similar subjects; surely they have nothing to do with mathematics. I collect and grade some written work early in the semester (usually in the form of a 20 minute quiz). Students are shocked when they read comments such as: "I cannot follow this," or "Where is the explanation?" or "This is not a sentence." followed often by the phrase "Not read further." When these comments are accompanied by large losses of credit, they begin to take my words and the handout as something with which they must cope. I ask all who have done poorly to come to my office for a conference and many others come to talk with me as well.

Usually they admit to not having read the handout. Some do not remember receiving it and others say that it was misplaced, so it pays to keep some spare copies on hand. Initial reactions range from incredulity to anger. Some sample encounters follow.

Comment: Dr. Henriksen, nobody ever asked me to write this way before.

Reply: That's too bad. The time to start writing clearly is now. How can I help you?

Query: I am sorry that you had trouble reading my explanations. Is it all right for me to go over them with you now?

Reply: Since you were told that you were expected to explain your work in writing, I cannot accept an oral explanation as a substitute and restore the credit you lost. On the other hand, I will be only too glad to tell you if your current idea of an explanation is a good one, and to help you find better ones.

Assertion: I got the right answer and anybody can understand what I am doing! (Usually said with irritation if not blatant anger.)

Reply: I am a professional mathematician and I can't understand what you have written. I see hardly any words, much less sentences. There are lots of symbols used without telling the reader what they denote. Would you like to go over some ways in which you might have explained your work clearly?

Assertion: I demand that you tell me how many points you are deducting for English and how many for mathematics! (Said with great anger.)

Reply: Since I could not understand your work, I cannot answer your question. Why don't you try to explain to me orally what you were doing and we can explore ways to express it in writing and see if it is correct.

The idea is to remain calm in the face of obvious attempts by students to remove your pants or to intimidate you with anger, and turn them in the direction of learning how to explain their work by writing in simple sentences. I did not always succeed in keeping cool. Once, while on a visit to another college and teaching first semester calculus, a student came to my office and proceeded to haggle over my grading of each question on the first hour exam. When I directed him to the handout, he told me that I would not be very popular if I kept on doing things like that. I could not hide my anger when I told him that I would penalize him even more in the future for poor explanations, and I would take care of my own problems with popularity. Thus, with this very bright (albeit insolent) student, I lost the chance to turn this into a constructive experience.

On the next quiz more students wrote in sentences, and there was a slow but steady improvement in the work of the students. Many resent having to change their ways, but bit by bit they fall into line, and it becomes much easier to read and understand their work. You have to stand fast as they test your determination repeatedly. Once they realize that you will not waiver, they start asking questions about what is a satisfactory mode of presentation, and the battle is won. It can take anywhere from half to two-thirds of a semester, but unless a student drops the course in anger, he or she does get in the habit of writing in sentences.

Those most likely to get angry at demands that they explain their work and write in sentences are at the extremes of mathematical ability. Some very bright students are used to "seeing" how to solve many problems in a *gestalt* without having to examine exactly what they are doing. I try to point out that this will not be possible forever and, hence, that they must learn how to explain what they are doing. It is particularly important that I convince such students that my writing demands are more than exercises in pedanticism. At the opposite extreme are the students who gave up on trying to understand mathematics. They had acquired certain skills in the past in a *gestalt*, and then moved to memorization when that no longer worked. Such students are very insecure and resent what they regard as yet another demand. When you can get them to listen, it is important to point out that good writing is a reflection of clear thinking, and clear thinking rather than memorization is the key to success in mathematics. I never accelerate my demands or spend class time haranguing my students on the virtues of good writing. It seems to be enough to go over briefly what it means to explain work and write in sentences, while never compromising in the face of complaints or abuse.

Good writing is a reflection of clear thinking, and clear thinking rather than memorization is the key to success in mathematics.

By now many who have been patient enough to read this far must be asking why it is worth accepting so much *sturm und drang* to attain such a modest goal? Why not spend the time needed to teach them how to write mathematics really well and assign a term paper? The answer is that any required course that I teach has such a crowded syllabus that I cannot spend the time needed to teach them how to write at such a high level without leaving out some of the topics our department is obligated to cover. It is difficult also to teach students how to write well if they regard writing in sentences as the equivalent of dressing up and going to church at Easter time. They can write in sentences, and doing so helps them break the habit of doing mathematics by throwing some symbols on a page and stirring them in hope of getting the "right" answer.

Some years ago a Pomona College senior enrolled in a pre-calculus course I was teaching that was normally taken by freshmen. He was a bright underachiever who had put off a mathematics requirement as long as he could. He ignored my handout and my request that he write in sentences over and over again. I called him into my office several times and asked him why he persisted in paying no attention to instructions, and said that I could not help him as long as his work consisted only of disconnected batches of symbols. When I discovered that his major was English, I lost patience and almost yelled at him, "You know what a sentence is! Why do you *refuse* to write in sentences?" I handed in a report of unsatisfactory work at mid-semester and I hoped secretly that he would drop the course and go away. Suddenly he began to explain his work and write clearly! By the end of the semester he had earned a grade of B and probably would have done better if it had lasted longer. He explained his long delay in trying to write anything comprehensible on his mathematics papers to his inability to associate it with anything but rote memorization. All that he had learned in the past had been on a *monkey-see-monkey-do* basis. This experience convinced me that the best favor I could do for my students was to insist they explain their work with the aid of complete English sentences!

This program always succeeds in that all of the students are writing in sentences well before the end of the semester, which makes it much easier to read and grade their work. In recent years I have included "answer only" questions on quizzes and hour exams, for which no partial credit is given and no work is read, to help convince students who think that they know the material, but get lower grades than they deserve because of my unreasonable demands, to realize that they are deluding themselves. While this helps a little, those that find that I will not choose among multiple answers to a single question or guess at the meaning of an ambiguous answer get just as upset as those who object to having to write in sentences. The upset might be less if similar demands were made in their other courses.

I have given up on trying to convince my colleagues to adopt this program. Some feel it unfair to demand that students write in sentences on examinations if their homework is not graded in the same way. Since homework is graded by undergraduates who are little better than answer checkers, and who themselves write poorly, this latter cannot be done. (I wonder if my colleagues would use this as an excuse if there were no students to grade homework.) Others plead a lack of time to engage in such a program, despite my offer to help and my claim that it saves time. I suspect that the real reason for the lack of willingness to insist that students explain their work is fear of complaints in an era in which popularity is equated with good teaching by so many college administrators.

In no way does this program take the place of the many much more serious attempts to teach students how to write substantial mathematical tomes. Indeed, when I teach advanced courses to students majoring in mathematics, a lot of my efforts are concerned with getting students to write in paragraphs. On those occasions when I wish to collect a term paper, I have to spend even more time teaching students how to write. The virtues of the program described above are that it is simpler to carry out, it reaches a wider audience, and it always seems to work.

I conclude with a sample handout on linear algebra from which the reader may construct one suitable for other courses.

Handout on Linear Algebra

Your grade in this course will depend almost exclusively on written work (homework, short quizzes, and examinations), so it is important that you learn how to communicate clearly in writing. Any work you submit for evaluation calls for an explanation of what you have done with the aid of complete, grammatically correct English sentences. (Symbols abbreviate English words or phrases and may be used as parts of sentences.) I will read exactly what you have written, and will make no attempt to deduce what you "really" mean or to supply missing steps or logical connectives. Any symbols you introduce that are not standard must also be explained or quantified. Make sure, also, that you supply an explicit answer to each problem you claim to solve.

In particular, I do not separate form from content. If I can't understand some part of your work, I will not struggle to read it, and your grade will suffer accordingly; even if you got the "right" answer.

Your explanations need not be lengthy to be clear. I conclude with some examples.

PROBLEM 1.

Find all vectors (x, y) in the plane such that $x + y = 0$ and $-3x + 2y = 1$.

Unsatisfactory Solution

$x + y = 0$

$-3x + 2y = 1$

$+3y + 2y = 1$

$y = 1/5, x = -1/5$

No explanation has been offered and the question has not been answered explicitly.

Satisfactory Solution

(1) $x + y = 0$

(2) $-3x + 2y = 1$

By (1), $x = -y$, whence by (2) $3y + 2y = 1$. So $y = 1/5$ and $x = -y = -1/5$. Thus the only such vector is $(x, y) = (-1/5, 1/5)$.

PROBLEM 2.

Suppose $A = \begin{bmatrix} 1 & 2 & -3 \\ 0 & 2 & 4 \\ -5 & 1 & 3 \end{bmatrix}$.

Evaluate the determinant $det\, A$.

Unsatisfactory Solution

$det\, A = 6 - 40 - 30 - 4 = -68$

The work has not been explained and there is no way for the reader to check its accuracy.

Satisfactory Solution

By the diagonal rule,

$$det\, A = 1 \cdot 2 \cdot 3 + 2 \cdot 4 \cdot (-5) - 3 \cdot 0 \cdot 1 - (-5)(2)(-3)$$
$$- 1 \cdot 4 \cdot 1 - 3 \cdot 0 \cdot 2$$
$$= 6 - 40 - 30 - 4$$
$$= -68.$$

PROBLEM 3.

State whether or not each of the following sets of functions is a subspace of the vector space V of all real-valued functions of a real variable.

(a) $S = \{y \in V : y'' + y = 0\}$.

(b) $T = \{y \in V : y'' + y = x\}$.

Unsatisfactory Solution

(a) $(af + bg)'' + (af + bg) = a(f'' + f) + b(g'' + g) = 0$.

Yes.

(b) $(2x)'' + 2x = 2x \neq x$.

No.

Symbols are used without definition. The work is cryptic. Neither question is answered explicitly.

Satisfactory Solution

(a) Suppose f and g are in S, and a and b are real numbers. By the sum and product rules for derivatives,

$$(af + bg)'' + (af + bg) = a(f'' + f) + b(g'' + g)$$
$$= a \cdot 0 + b \cdot 0 = 0.$$

Hence S is a subspace of V.

(b) Let $h(x) = x$. Then

$$h''(x) + h(x) = 0 + x = x,$$

so h is in T. But

$$2h''(x) + 2h(x) = 2x \neq x,$$

so $2h$ is not in T. Thus T is not a subspace of V. [Note that a subset K of V is a subspace of V if and only if $a, b \in R$, and $f, g \in K$ imply that $(af + bg) \in K$]

Three R's for Mathematics Papers: 'Riting, Refereeing, and Rewriting

Thomas Q. Sibley
St. John's University, Collegeville, Minnesota

Major papers each year reprove their educational worth in my upper division mathematics courses. The students' stereotypical groans and even apprehension in early September annually turn into hard work and considerable enthusiasm in November, resulting in a lot of valuable learning and many good papers in December. I require all students to critique two other papers and all of them to rewrite their own papers based on these critiques and my comments. This refereeing and rewriting, I believe, are the most important factors in the high quality of the papers. This article describes the procedures I have developed for these papers, together with some perspective on my experience.

My students have investigated a great variety of topics, including fractals, pedagogical issues in secondary mathematics education, computer graphics, the group theory of change ringing, knot theory, and the interplay of mathematics and Islamic art. The students, with some professorial suggestions, generally pick topics matching their abilities and wide interests, which enables me to learn about at least one new topic each year. They learn much more than subject content and expository writing skills—too few of them have previously faced the opportunity to learn mathematics on their own and fewer still have searched the library stacks and journals for information on a mathematical topic.

My chief reward from these papers is seeing them learning mathematics and learning how to learn. Next comes the pleasure of reading interesting papers, especially on topics new to me. My six years of experience with assigned papers suggest two added benefits. The more mathematically ambitious students come more frequently to consult with me, giving me a chance to know students who sometimes have had no need to seek help. Further, when I write letters of recommendation, I find it much easier to write something special about the students who have written papers for me. I have a broader understanding of these students, as well as a specific context in which to discuss their strengths and weaknesses.

The student refereeing of others' papers and the rewriting I require each student to do, I believe, greatly increase both the quality of the finished papers and the pride the students take in their work. While I am rather hectically reading all of their papers and commenting on them in a week, each student is providing feedback on two of the papers. Then, armed with three critiques—mine and two students'—each author has another week to polish or repair (or occasionally salvage) the paper.

The students benefit in multiple ways from the refereeing. First, they become more sensitive to the need for clear writing as they scrutinize their peers' prose. Second, they learn the content of other students' papers. Third, the student readers often make valuable comments on aspects I never noticed. (This is a relief to me, since I no longer feel that I have to be the sole authority on writing.) Fourth, the student readers provide a supportive audience; they definitely take to heart my admonition to give both compliments and constructive criticism. Every paper benefits noticeably from rewriting, even the ones to which I was ready to assign an "A" on the first reading. The students often achieve a deeper understanding of their topic due to the critiques, and they definitely improve their mathematical prose.

On the first day of class, I provide three pages of information about the papers, along with the syllabus for the course. I tell them the due dates for choosing their topics, turning in the paper, refereeing and finishing the rewriting. I also include a description of the format, a warning about plagiarism, a long list of possible topics, and a description of the audience and depth of the papers. Eight to fifteen pages provide ample room for their topics. (Few students have trouble writing eight pages.) I ask them to write their papers at the level of the other students in the class, eliminating repetition of material they should all know. On the other hand, I expect them to delve deeply enough to ensure that all of their intended audience will learn something significant.

The student refereeing of others' papers and the rewriting I require each student to do, I believe, greatly increase both the quality of the finished papers and the pride the students take in their work.

I require them to have my approval for their topic a month before it is due. Not only does this shrink the number of inappropriate topics, it spurs procrastinators and allows me to suggest sources and focus for the topics before they start writing. The suggestions on sources supplement the extensive annotated bibliography I provide on the first day of the class.

The flurry of activity starts for me on the due date. I make it quite clear that I have to be inflexible about the due date because of the strict schedule of refereeing and rewriting we need to follow. I assign referees based on the difficulty of the topic, the mathematical ability of the referee, and the writing abilities of the author and the referee, as best as I know them. Then I parcel out copies of the papers to the student referees and do my best to read all of them carefully and comment extensively. I try to focus on major points of exposition and mathematical content, especially the correctness of the mathematics. I do make it clear that I deem poor English usage unacceptable; and I hold to it, marking the original paper and, if necessary, refusing to grade the rewritten paper until it is acceptably written. In every paper I seek aspects to compliment and some significant improvements to suggest. Although a week never feels long enough to accomplish this, I know I can not afford any more time, so I hold myself to that deadline as strictly as I hold the student referees. At this time I jot in my grade book provisional grades for the papers, which makes the actual grading much quicker.

A week after the due date, I collect all the referees' critiques, make copies for me to read, and give their critiques and mine to the authors. I ask the referees to assess the papers in several ways. First of all, they need to provide an overview of the organization, interest, and style. Next they should discuss the mathematical content, focusing on the clarity and level of the presentation for them, the intended audience. Finally, I ask them to comment in greater detail on the writing. Their critiques sometimes suggest pointers for my own future feedback to students.

The rewritten papers generally receive from me only short comments and the grade. Since the students are finished with their papers, there seems little point in extensive responses. I look especially for the changes they have made as I reassess the grade I assign.

I keep a copy of each paper, and I have consulted a number of them. I believe these papers provide a worthwhile resource, particularly for future high school mathematics teachers. With small classes, I give each student a copy of all of the papers on the last day of class. Two years ago, the geometry class had twenty-seven students, too many, I judged, for the department to pay for all of the copying. Many of the students willingly paid the duplicating cost for a set of the papers.

Major papers in mathematics classes entail two drawbacks: time and time. First of all, students writing, critiquing, and rewriting papers clearly have less time to spend learning standard material, so the content of the course needs modification. In geometry and the second semester of abstract algebra, where I have assigned papers, the changes caused by the papers seem to me only adjustments, leading to different but equally valuable learning. However, I feel that linear algebra, multivariable calculus, and other mathematics courses with a large amount of specific, standard content have little room for papers.

The second drawback concerns my time—the great quantities of time to thoughtfully read, comment on, and grade a stack of papers. I know of no short cut for this. Nevertheless, each year, including the year with twenty-seven geometry students, the papers I read amply justify the hours I spend.

I believe the opportunity to research, write, and rewrite about a topic is one of the best educational assignments I give my students and one of the most rewarding for me personally.

For courses which have sufficient flexibility in their content, expository papers can engage each student in an individualized learning experience, rewarding and challenging to average students and to outstanding students alike. The required refereeing and rewriting dramatically enhance the students' final results. I believe the opportunity to research, write, and rewrite about a topic is one of the best educational assignments I give my students and one of the most rewarding for me personally.

WHAT DO STUDENTS SAY?

Attempting Mathematics in a Meaningless Language

Martha B. Burton
University of New Hampshire, Durham, New Hampshire

A colleague recently recounted an undergraduate's complaint about his history-of-mathematics textbook. The otherwise passable text, said the student, was too frequently interrupted by objectionable parades of *symbols* across the page. This student had proceeded with equanimity through pages of English words, which are nothing if not symbols themselves. Thus the problem was not really one of dealing with a symbolic language, but of confronting symbols of an unknown language—empty symbols, symbols with missing referents.

The minimal characteristic of a symbol is that it points beyond itself to something else. A phrase in our natural language has a referent in our own experience: "jet airplane" is a real symbol because we know what a jet airplane is. "Slithy tove" is an empty symbol, having no referent in our experience.

Beginning undergraduates in mathematics need to use the symbol-system of elementary algebra as a language, but it is common for them to experience phrases of the algebraic language as symbols without referents. The student who describes the meaning of the symbol x as "x ... just x" is not apt to find any referent for the symbolic phrase $1/\sqrt{x}$. Even more common is to experience an algebraic symbol as empty in the sense of referring only to other algebraic phrases, so that the referent of $1/\sqrt{x}$ might be simply $\sqrt{x}/x$. If this is a referent, it is of the most trivial and circular kind: according to this understanding algebraic phrases can have no meanings, but only synonyms.

Yet no real language is so impoverished as to describe only its own words. Language has to be *about* something to be real, and even students who are mute in the algebraic language usually believe that it functions as a language. It is to be seen in books and journals, so that evidently it supports at least some kind of communication between at least some sorts of people. Do mathematicians at supper converse in this strange manner? The students' question is not entirely frivolous, for they usually follow it with the *right* question about a language: What do you talk about?

The students whose mathematical activities began these speculations are first-year calculus students at the University of New Hampshire, particularly those who visit a Mathematics Department facility called the Mathematics Center. Some are doing required first-semester remediation in elementary algebra and trigonometry; some are working on projects in numerical integration; some drop in with calculus questions. They afford the Mathematics Center's staff a privileged view of the ways in which large numbers of students address problems in mathematics. Mathematical misconceptions that might otherwise be put down to individual error sometimes command an observer's attention just because so many students display them.

A large part of our students' difficulty with calculus would appear to be language-based. They seem to use their algebraic language almost exclusively in its calculational or syntactic aspect, finding its descriptive, meaningful aspect inaccessible. Thus, for instance, they can complete word problem solutions—once the meaning of the problem has been distilled into an algebraic sentence. They eventually wade through their algebra remediation—provided they concentrate on syntactic examples, those having to do with solving algebraic sentences or rewriting algebraic phrases. They find limits exercises in calculus baffling, since these present only algebraic phrases whose values are to be estimated, whereas the student prefers algebraic sentences which might be solved by grammatical rearrangement. In all these situations the students are willing and able to manipulate the syntax of algebraic sentences, but cannot find or express meaning in the algebraic language.

A frustrating and finally convincing experience in the Mathematics Center began with a certain hint included in a numerical integration problem set. This one-line hint was clearly stated in algebraic terms; yet not only was it incomprehensible to most students, its crucial part was actually invisible to many of them. We will shortly see how this careful attempt to communicate information to students in the algebraic language met with utter failure.

Algebra as a Subset of English

To the extent that students' difficulties are related to their misuse of the algebraic language, it is worth looking carefully not only at student use of the language but at the characteristics of the language itself. It is the assumption throughout this paper that the system of algebraic symbols is not an entirely separate language. Rather, in its descriptive aspect at least, it is a telegraphically-written subset of our natural, spoken language. Sentences written in the algebraic language are to be understood in the natural language: variables, even those denoted by letters, are to be understood as quantity *names*. The calculus instructor who writes "$w = fd$" on a blackboard is simultaneously uttering some English sentence containing the nouns "force," "work," and "distance." Unfortunately, many calculus students who copy "$w = fd$" from the blackboard into their notebooks will thereafter read the sentence literally, with no mental reference to English nouns.

Assuming that the algebraic language is a subset of English, we should note that the inclusion is proper. There is, for instance, no algebraic equivalent to "The dog needs a bath." So a reasonable first question is: which English sentences have algebraic equivalents?

First-language acquisition by young children is a remarkable process, not completely understood and of wide interest among linguists and psychologists. In this connection one linguist writes:

> For each English active sentence with a certain sort of transitive verb (including *eat, bite, catch* ... etc., but excluding *cost, weigh* ... etc.) there exists a corresponding passive sentence with the same verb ... [4]

From our point of view, the striking thing is that the example's excluded verbs are just such as might occur in the word problems of elementary algebra. Hints should be taken wherever they are found, and this one suggests that only certain verbs will occur in English sentences that have algebraic equivalents.

The English sentences that have algebraic translations are those having to do with quantities. They have nouns that represent quantifiable entities, and often their verbs—like *cost* and *weigh* in the example—correspond to measure functions. These words, appearing as verbs in the English sentence, will instead appear as part of a noun phrase in its algebraic version. For instance, corresponding to the verb "weigh" is the measure function $weight_{of}$, so that the English sentence "The dog weighs 75 pounds" has for its algebraic equivalent "$weight_{of}$(dog) is 75."

By the time the encoding of an English sentence into the algebraic language is complete, all its nouns will have been treated in this way. Each will have had applied to it some measure function whose value is a real number. The English sentence's verb may appear as a measuring modifier in an algebraic translation.

Since the noun phrases as they appear in the algebraic version are values of measure functions, the verbs that can occur in the algebraic translation correspond to relational operators. Basically only two verbs will appear in algebraic sentences: they are *is* and *exceeds*. Combinations and rearrangements of these, such as "less than or equal to," are grammatical variations of these two verbs.

Once the English sentence has been encoded algebraically, it has become an algebraic statement relating real numbers. It may have become, for example, the desired equation that summarizes some given problem. Rewriting the algebraic sentence according to the syntax of the algebraic language is a process that amounts to solving the equation, for the grammatical transformations appropriate to the algebraic language are simply calls on the axiom system of the real numbers.

Some Semantic and Syntactic Peculiarities of the Algebraic Language

Every language has a *semantic* component—its descriptive aspect, the part of the language that carries meanings—and a *syntactic*, or grammatical, component. In the case of the algebraic language there is an asymmetry between the semantic and syntactic components that is clear, but worth explicit mention. The language of elementary algebra, being a shortened version of our natural language, borrows the natural language's semantics. The semantics of the algebraic language is not fixed: w might mean "work done" or "weight" or something else entirely. Nevertheless, whatever specific meaning we assign to w will be a meaning from our natural language.

By contrast, the syntax of the algebraic language is that of the real numbers. Sentences in the algebraic language are subject to grammatical transformations which reflect the arithmetic properties of the real numbers. Indeed, the algebraic language is invoked in problem solving precisely in anticipation of grammatical transformations on an algebraic phrase or sentence.

The syntactic power of the algebraic language, providing as it does the means of solving problems, is the source of its importance. Unlike the grammar of a child's first language, the grammar of the algebraic language is explicitly *taught* to students. Most of a course in elementary algebra represents instruction in its grammar. It is hardly surprising that students use the language of algebraic symbols in a primarily grammatical, algorithmic spirit.

The algebraic language's peculiar divergence of semantics and syntax, with its semantics reflecting the natural language and its syntax deriving from the real numbers, sets up an interesting trap for the student who must find meaning in the language. To see where this bind is, we must consider what a meaning might be, and in particular what it might be in the algebraic language.

Meanings of Algebraic Sentences

There is great difficulty in specifying just what meanings are, and we will have to be satisfied with approximate and philosophically fuzzy descriptions. For instance, even if we could confidently specify the meaning of a simple declarative sentence like "The grass is green,"

we might still be hard put to describe the meanings of related queries ("Is the grass green yet?"), commands ("Green up, grass!") or phrases ("the green grass").The meaning of a simple declarative sentence—or, at least, something very close to its meaning—is often taken to be its *intension*. An intension is a function mapping "possible worlds" onto the set $\{T, F\}$ of truth values of sentences. Each of us at each moment of our lives inhabits a different particular world, so that by a possible world we should understand the totality of our views, knowledge, and experience at some instant. The intension of the sentence "The grass is green" is a function which, applied to a person's own world, produces one of the values $\{T, F\}$. It is the means by which we verify that sentence in our own experience.

A different example may make this notion more appealing. Suppose we are computing values of the expression $3x + 4y$, and we have integer values of x and y stored in a symbol table. How shall we describe the meaning of the expression $3x + 4y$? It should be distinct from any one of its values, which are integers, for it depends on the stored values of x and y. It makes sense to take the expression's meaning to be a function, mapping possible symbol tables into the set of the expression's possible values. Now if we take our own knowledge and experience to be a kind of grand "symbol table" and the values of sentences to be their truth values, we arrive at something very like the idea of the intension mapping.

If this definition of meaning seems scholastic and ethereal, imagine our surprise when we saw calculus students in the Mathematics Center acting it out. What follows is a more complete description of the failed hint, earlier mentioned, that was included in a problem set on numerical integration. The thing to notice is that when students were presented with an algebraic statement meant only to convey information to them, their response was to attempt verification of the sentence. Since the means of verification in the algebraic language are ready to hand, and are syntactic in nature, the very attempt to access the statement's meaning detoured the students into its syntax.

The hint was intended to impress upon the students that error estimation in Simpson's rule requires use of the integrand's fourth derivative, whose form may be extremely awkward. At the same time the students were offered an upper bound for the unwieldy derivative:

Hint:

$$|f^{(4)}(x)|$$
$$= \frac{|(1 + \cos x)(1 - 5\cos x - 2\cos^2 x) - 3\sin x(5 - \cos x)|}{(1 + \cos x)^2}.$$
$$\leq 34$$

We soon knew that we had done the wrong thing. Repeatedly students told us they couldn't understand the hint. Many students immediately set about calculating an upper bound for the ungainly expression in the hint. Very few of them took any note of the number 34 offered, and a surprising number told us they had not even been able to see it. Confronted with information written in the algebraic language, the students instead tried to compute something with it: in attempting to understand the hint they were diverted from the language's descriptive aspect into its algorithmic one.

Eventually, of course, the students short-circuited their problem. The student grapevine is very efficient, and the word went out, "Don't even think about it, M is 34." A subsequent edition of this problem set contains the same hint written partly in English, which has settled the difficulty.

Producing Algebraic Sentences:
Some Ways that Work, and Some that Don't

Inability to extract meaning from an algebraic sentence is one part of our students' difficulty with this essential language. Even more obvious is the converse problem of expressing meaning in the language. This is apparent in their reluctance to attack word problems, where the missing skill is that of summarizing the meaning of the problem in algebraic language. Let us consider first a pair of strategies that do *not* help students produce algebraic descriptions of problems.

The first of these is advice to the student to read the problem carefully. This good advice is redundant, for the cause of failure with a problem is apt not to be careless reading of it. Reading students' incorrect attempts at word problem solutions will often make it plain that they have indeed understood the problem requirements, but have not been able to summarize them in an algebraic sentence.

The other trouble with advice about careful reading is that word problems in elementary algebra and calculus are often chock-full of data. As with a tangled skein of yarn, finding the loose end is more profitable than carefully contemplating the whole tangle. What the student needs is not so much careful attention to the whole of the problem as a way to find the particular piece of data or point of view that will start the problem unraveling.

A second piece of advice commonly given to students is to begin a word problem by giving every quantity in it a letter-variable name. Students seem universally to have accepted this advice; they are everywhere to be seen dutifully initiating problems with glossaries like "let $x = \ldots$, let $y = \ldots$." There are several things wrong with this approach.

What the student needs is not so much careful attention to the whole of the problem as a way to find the particular piece of data or point of view that will start the problem unraveling.

For instance, the names "x" and "y" convey no meanings. Students use literal variables because they think it is the correct thing to do, yet their use starves the student in the midst of semantic plenty. In its descriptive aspect, the algebraic language is part of English, with English nouns available for use as variables. It is pointless to begin a problem with the statement "let $x =$ width of pasture" when the English word "width" would have been a more meaningful variable, requiring no glossary entry at all. The construction of a variable glossary for a word problem is a preliminary task for the student that yields no further insight into the problem. At best it produces a correspondence between English noun phrases and their algebraic equivalents, giving the student something extra to remember.

But the chief reason that glossary-building is nonproductive is that it provides references only to the sentence's noun phrases. With only the nouns in hand, the student will still not know what to say about them. The trouble is that the verb is missing. Until the problem statement has a verb, no collection of noun phrases will lead the student to the desired equation.

Verbs turn phrases into communications. In the construction of an algebraic sentence, the crucial step is to combine noun phrases meaningfully with a verb. The traditional method of constructing an algebraic sentence is to do it wholly within the algebraic language. One names the variables, collects algebraic noun phrases and then (somehow) assembles them with an appropriate combination of the verbs "is" and "exceeds." If this approach were productive, students would be much more successful with word problems than they are.

A better method, we believe, is to assemble the whole sentence in English first. The student then has the advantage of using the familiar natural language for thinking about the problem's summary, and the English summary sentence will emerge already equipped with its verb. Naming the problem variables is deferred until the summary sentence is written.

Verbs turn phrases into communications.

Students need some permission and guidance in beginning problems in this way, but oddly enough they do not need help in supplying suitable variable names. Providing invented names for things is an activity humans, even very young children, are known to do without instruction. [2] In fact, one has only to listen to their everyday conversation to hear students produce variable names: a recent favorite has been "Well . . . *whatever*." It does no harm, therefore, to omit the formal construction of a variable glossary, proceed directly to the stage of supplying a sentence, and allow a deferred variable-naming process to occur naturally.

Setting Up Word Problems

To see how problem-solving algorithms based on these considerations might differ from those that are usually offered to students, we outline the Mathematics Center's directions for organizing maximum-minimum problems in calculus.

Students are first asked to be sure that the problem really is a maximum-minimum problem, and to write down in English the name of the quantity to be optimized. This has the effect of encouraging the student to read the problem carefully, yet offers focus to the reading. The student is also told to illustrate the problem with a sketch.

The student is next asked to write a sentence describing the target quantity in terms of other problem quantities. The sentence can be brief, but it should be in English: "*Area* is length times width," or "*Total Area* is area of circle plus area of square," or "*Profit* is income less total cost."

Finally, if the target quantity depends on more than one other quantity, we ask the student to return to the problem for extra information relating these "competing independent" variables. The advantage of describing the target quantity first, and only afterwards considering side conditions relating other quantities, is to keep the problem's main statement central in the imagination. Without some kind of hierarchy among the problem's relationships, the student is apt to be diverted by relationships that are valid but do not lead to a solution.

The statement-first, top-down strategy can quickly produce the problem's central equation, complete with a suitable independent variable. For instance, the following maximum-minimum problem was taken from an old calculus hour examination. The calculus lecturers asked that this sample exam be posted in the Mathematics Center, where it naturally provoked student interest and discussion.

Example. Citrus Hill can expect an average yield of 40 bushels of oranges per tree when it plants 20 trees on an acre of one of its fields. A researcher has learned that each time an additional tree is planted on an acre, the yield per tree decreases by one-half bushel. How many trees should be planted per acre in order to produce a maximum yield?

On a recent afternoon three calculus students and one of the Mathematics Center student workers, a senior mathematics major, spent ten minutes considering how to set up an equation for this problem.

The focus of their discussion was on the quantity to be chosen for the independent variable. Because the problem asked "How many trees ... ?" the students proposed making $number_{of\ trees}$ the independent variable; but then they were confused by the base condition of the 20 initial trees. The student worker pointed out that 20 trees correspond to 40 bushels per tree, 21 trees to 39.5 bushels, 22 trees to 39 bushels, etc. Considering a small table of this data led the students to an appropriate choice of independent variable.

By contrast, notice how naturally an equation shows itself when we proceed top-down, beginning in English with a description of the target variable $Yield$ which is to be maximized:

$$Yield \text{ is } number_{of\ trees} \text{ times } bushels_{per\ tree}.$$

Because $Yield$ is not yet described in terms of only one independent quantity, we recursively describe the successor expressions $number_{of\ trees}$ and $bushels_{per\ tree}$:

$$number_{of\ trees} \text{ is } 20 + extra_{trees}$$

$$bushels_{per\ tree} \text{ is } 40 - extra_{trees}/2$$

The description of $Yield$ is now

$$Yield = (20 + extra_{trees})(40 - extra_{trees}/2)$$

which the student has probably already abbreviated to

$$y = (20 + x)(40 - x/2).$$

Notice that not only has the problem's central equation emerged naturally, but so has a suitable independent variable.

Maximum-minimum word problems in calculus are more easily set up than are the word problems of elementary algebra. They are more uniform in type, and the quantity to be optimized is usually easy to spot. The word problems found in elementary algebra texts are more varied in kind, so that authors often separate them into types according to the standard formula whose use is anticipated.

To use a statement-first scheme in the solution of a word problem from general elementary algebra, the English sentence should summarize the relationship among quantities after the problem's requirements are met. For instance:

Example. Sonya, Henri, and Felix are grading a stack of calculus quizzes. Sonya could grade the whole stack alone in 2 hours, Henri in 3 hours, and Felix in 4 hours. How long will it take them to do the work together?

Students may remember that this is a standard sort of problem, and that a certain standard trick makes it work out easily. But it can be begun with a sentence summary:

Sonya's portion + Henri's portion + Felix's portion = 1 whole stack of papers.

Each person's portion can be re-described. Sonya's share, for instance, will be 1/2 stack per hour, multiplied by the time the committee spends reading the papers. Rewriting each reader's share in this way leads immediately to an equation in the one variable $time_{spent}$.

Visualizing Problems

Pictorial imagination is a crucial part of mathematical problem-solving activity, and one which many students fail to use effectively. It would be misleading to classify this kind of imagination as "nonverbal" in the sense of being something quite divorced from language. On the contrary, it is worth considering the differing pictorial resources of the two languages calculus students must use, and the situations in which they are apt (or not apt) to picture their problems.

Pictorial imagination is a crucial part of mathematical problem-solving activity, and one which many students fail to use effectively.

The natural language first acquired by young children is rich in the names of familiar objects. Indeed, first books in the nursery are usually simple collections of illustrations of named objects. The profuseness of illustration in small children's storybooks, together with the low proportion of text to illustration, make it clear that the text alone cannot convey the meanings intended. Not only are the pictures an essential part of the stories, but the visual images are actually part of the language itself.

As readers mature, written text supplants much of the illustration in reading material. It is not to be supposed that older readers have ceased to enjoy pictures, but rather that the expanding language acquires more and more of its own visual power. Popular adult paperback mysteries, even those we should describe as graphic, contain no line drawings. They are not needed, because the language itself is visually sufficient. Poetic language is a more extreme example. It can be so visual that added illustration would be not only superfluous but an annoying impediment: "Teintés d'azur, glacés de rose, lamés d'or" is not amenable to illustration.

There is no natural visual world appropriate to the algebraic language. A specifically invented environment, the Cartesian coordinate system, provides the language's visual component. The visual aspects of the algebraic language as they are illustrated in analytic geometry differ in two important ways from the natural world and its richly visual language.

One important difference is in the visual power of the languages. Even with its invented visual facility adjoined, the terse and pithy algebraic language does not enjoy the visual richness of natural language. First-year calculus books are rich in necessary supplemental illustrative figures. As with the pictures in children's books, the figures in a calculus text are not extraneous but are rather part of the language of exposition. The algebraic language itself, at least in its semantic aspects, can require accompanying illustration for completeness.

Another difference is in the syntactic parts of the language that can be pictured. In natural language, nouns and noun phrases are illustrated: "the boy in blue," "the mountains at sunrise," "two spheres." On the other hand, in the coordinate geometry of, say, two dimensions the primitive recorded objects are ordered pairs of numbers. As the recorded pairs have coordinates with some relation to each other, it is essentially algebraic sentences, not algebraic noun phrases, that are displayed. The sentence $y = x^2 + 3$ can be illustrated, but not the phrase $x^2 + 3$.

Regions in the coordinate plane are visible only in the sense that we display the sentences that form their boundaries. With a distance measure applied to points on the bounding graphs, we can ascribe geometric notions like height and breadth to their enclosed planar regions. But the regions themselves are still constructs of the imagination, and it can be nonintuitive to students that regions in this entirely artificial coordinate plane are described and measured as if they had natural-world reality. A certain level of mathematical indoctrination is required before the Cartesian coordinate system achieves the reality of a landscape in the natural world.

First-year students often have trouble visualizing geometric applications of integration such as areas, volumes of revolution, and centroids. With surprising frequency students attempt these problems without any sketches at all, simply inserting likely-looking algebraic phrases into standard formulas that they have memorized. Thus they delete from their algebraic language its necessary added visual component, with results that are usually disastrous.

It should not be imagined that students use memorized formulas in this unilluminated way because they cannot picture the requirements of problems. Many of them are studying physics and engineering, where presumably they are used to problems that must be visualized. Indeed we have watched them do their trigonometry remediation and seen their many sketches showing flagpoles atop buildings, forest rangers viewing fires, and brave gunners aiming cannons at the castle of the evil Count. Even when they can go no further with a word problem, students are apt to illustrate it.

There seems to be a difference between those problems for which our students consent to make sketches, and those which they begin without any illustration. Students do generally make sketches for problems that originate in their natural language. They sketch views of their natural world, or of some imagined variant of it. They correctly see illustration as one of the stages in converting a natural-language problem for algebraic resolution.

It is chiefly when the problem both originates and is resolved within an algebraic framework that students slight the task of sketching. This should not be surprising. The sketch must be done within the algebraic language's less familiar visual environment, and drawing it is a curve-sketching task in analytic geometry. The techniques for this are probably not yet part of the student's easy competence.

Geometric applications such as volumes-of-revolution problems usually originate in the context of analytic geometry and in the language of algebra. Not only are these problems subject to the difficulties just noted, but they reach the student with associated standard formulas already provided. With both the problem and its resolving formula written in the algebraic language, the student finds it reasonable to proceed by transcribing algebraic fragments from the problem into the formula.

Encouraging Visualization

Considering the visual properties of their languages suggests some general ways of fostering mathematical visualization by our students. We need to encourage students, both directly and indirectly, to make and interpret sketches in order to do their homework problems. Sometimes the most direct route is appropriate: students need to be directly told to do this. But we also need other ways to interrupt the short circuit from algebraically described problems to algebraic language formulas, in which the student merely transfers phrases from problem to formula.

One simple approach is to use fragments of the more visual natural language in both summary formulas and their accompanying sketches. It may be less elegant to describe an element of volume as $2\pi \cdot radius \cdot height \cdot thickness$ than to describe it as $2\pi p(x)h(x)dx$ or $2\pi p(y)h(y)dy$, but it's a lot clearer. The natural language description leaves it to the student to decide whether the

volume element's "thickness" corresponds to a differential in x or y, and then to describe "radius" and "height" in terms of that same variable. This procedure, admittedly wordy, nevertheless shows the student the decisions that actually have to be made in setting up the integral. By contrast the fragment $p(x)$ in the standard algebraic language neatly but tacitly conveys that since the variable of integration is x, the radius of the shell will have to be described in terms of that variable alone. The message may simply be too tacit for the student reader, who often assumes that $p(x)$ represents some algebraic expression already specified in the problem. Good students have told us that they were confused about what to "put down for $p(x)$."

But we also need other ways to interrupt the short circuit from algebraically described problems to algebraic language formulas, in which the student merely transfers phrases from problem to formula.

Students find it important to memorize formulas. It is an activity we can count on them to continue. It will thus be advantageous to encapsulate more meaning into the formulas they are going to memorize anyway, and we have just suggested that one way to do this is to use natural language fragments in their standard formulas. Another way would be to place a relevant sketch with each formula. This suggestion may sound supremely redundant, considering the multitude of sketches already in calculus texts. However, it has become standard practice for texts to appear with important facts and formulas highlighted on the page in boxes of some second color. These are the things the students memorize, and if a formula is in a colored box then a sketch of its application ought to be there also.

Both these suggested remedies have the flavor of a quick fix or patch; and to be sure, there are times when a well-chosen patch is needed and appropriate. But another view of our possible responses begins when we compare the sketches the students make with those they omit. They will avoid the curve-sketching involved in illustrating a volumes of revolution problem, where they would not hesitate to sketch the natural-world dimensions involved in a maximum-minimum problem.

The reluctance of students to make sketches within the coordinate plane's geometry probably has no quick-fix solution. To the extent that students have trouble because they cannot visualize geometric constructs in the coordinate plane, the remedy is more time and effort invested in the teaching of analytic geometry.

Names and their Symbolic Referents

We have considered several aspects of the algebraic language that calculus students find necessary to use but difficult to use meaningfully. The central thread of this troubled web is their attempt to use algebraic names divorced from their symbolic referents.

Earlier we said that the minimal property of a symbol was its ability to point beyond itself to something else. A stronger property would be for the symbol to enjoy some participation in the thing to which it points. A national flag participates symbolically in the dignity of its nation, making it a symbol in the strong sense. By contrast, the uniform yellow color of school buses is a weak symbol, simply a convenient way of making the vehicle obvious and easy to recognize. People are offended to see the flag used in inappropriate ways, whereas they would be merely surprised to see a car painted school-bus yellow. The stronger the symbol, the more troublesome is its use apart from its proper referent.

Names in general have historically been seen as important symbols, sometimes having been extremely strongly linked to the persons or things named. The use of a name without respect for its symbolic linkage has been found offensive in different ways in various circumstances and cultures.

It is worth remembering that mathematical naming is part of a wider and older practice, long associated with peculiar symbolic powers and hazards. An algebraic name, once attached to a number, participates in the algebraic syntax exactly as its referent number would: the strength of its symbolic connection with the named quantity is what makes the name so useful. By the same token, that very symbolic strength makes an empty use of an algebraic name worse than useless. For our students the vacuous use of algebraic names is ruinous, reducing their understanding of mathematics to meaningless and formal exercises on sets of letters.

REFERENCES

1. Martha B. Burton, "A Linguistic Basis for Student Difficulties with Algebra," *For the Learning of Mathematics*, 8 (1988): 2–7.

2. E. V. Clark, "Convention and Contrast in Acquiring the Lexicon," in Thomas B. Seiler and Wolfgang Wannenmacher (eds.), *Concept Development and the Development of Word Meaning*, Springer-Verlag, Berlin, 1983.

3. Janet Dean Fodor, *Semantics: Theories of Meaning in Generative Grammar*, Harper & Row, New York, 1977.

4. Jerry A. Fodor, "How to Learn to Talk: Some Simple Ways," in Frank Smith and George A. Miller (eds.), *The Genesis of Language: A Psycholinguistic Approach*, MIT Press, Cambridge, Massachusetts, 1966.

Using Expressive Writing to Support Mathematics Instruction: Benefits for the Student, Teacher, and Classroom

Barbara J. Rose
Roberts Wesleyan College, Rochester, New York

Since I have come to believe that expressive writing can enhance the learning of mathematics I have been conducting my own research in college-level courses at Roberts Wesleyan College, a small liberal arts college in Rochester, New York. Some of that research focuses on my own experiences with a particular class of sixteen students enrolled in a three semester credit calculus course, the second of a two-part math sequence primarily for business majors.

As teacher of the course, I designed and implemented writing activities which were integrated with the regular teaching of the course and offered students a variety of opportunities for expressive writing: spontaneous dialogue journal writing, autobiographical narratives, and focused in-class writing.

Besides collecting and coding the writing produced by the students throughout the course, I carefully monitored the experience by gathering further information about the students' backgrounds, their explicit evaluations of the experience (through both in-depth, written evaluations and interviews), and my own field notes. I tried to use a variety of qualitative methods to analyze the data.

In this paper, I will describe the writing activities and discuss the benefits of writing to the student as writer, to the teacher as reader, and to the classroom situation as a result of this kind of student-teacher interaction.

Writing Activities

SPONTANEOUS DIALOGUE JOURNAL WRITING   On the first day of class, I gave each student a spiral-bound notebook in which to keep all writing assignments; a sheet containing the following rationale and procedures for their journals was distributed and discussed:

> A journal is a place where you can write to and for yourself about mathematics. Through writing, you can reflect upon and express feelings about mathematics and the course that affect learning, both positive and negative. You can also use the journal as a vehicle to "write to learn," by thinking on paper about mathematical concepts and processes.

> Journal writing is "freewriting;" that is, it is intended to be unedited, uncensored writing, and differs from the writing "product" you ordinarily submit for a grade. There is no such thing as a "wrong answer" journal entry. Simply let your pen take off and see what happens!

> Use your journal to think on paper about mathematical concepts; summarize the text or class discussion; write definitions, formulas, or processes in your own words; raise questions; write your way through a problem you can't solve; or record your feelings about the course, specific material covered, the teacher, homework, the book, tests, use of class time, or the nature of mathematics.

> Although the journal is intended primarily as a means for you to write to yourself, it can also serve as a dialogue with the teacher, giving important information to the teacher concerning how you feel about the course and material, so as to enable better use of class time and to improve instruction.

> Write as many entries per week as class hours for that week (usually three) on the right-hand pages of your writing book, with each entry dated, written in ink, at least one page long, and with no more than one entry per page. Bring your journals to class every period since we will also be writing on the left-hand pages of the journal during class. The journals will be collected at random, unannounced, and returned to you during the next class with comments. The grading will be based on maintaining the frequency and volume of the writing, not on mechanics or content.

A journal is a place where you can write to and for yourself about mathematics.

I intentionally gave spiral-bound notebooks because they preserved the chronological and developmental *record* over the duration of the course and kept individual assignments from being lost. Students were required to write in ink because it symbolized that the writing was intended in an exploratory way, without regard to mechanics or form; there were no "right" answers, and all writing was important and should be preserved to reveal current feelings and thoughts as well as to provide an ongoing and permanent record of what was felt and thought.

The journals were spontaneous and unstructured in the sense that students were free to write about anything pertaining to mathematics and the course. My past experiences with journals, however, revealed that students often complained about not knowing what to write about in a mathematics journal. Consequently, I gave students a list of 34 possible journal topics to be used only if needed or desired, such as: "Summarize in your own words the most important thing you learned in class today." "How do you read the textbook?" "What do you expect from the teacher?" "Describe how you do a particular process or problem." "Is mathematics invented or discovered?" "How should we use class time to best advantage?"

There were, however, three instances during the semester when I structured the journals. The first occurred at the beginning of the course when students were instructed to write autobiographical narratives and a mathematical image (the entry was to begin with "Math is like . . ."). The second was a response to the unit tests, with the following questions as a guide:

Before the test:
    How do I feel about taking this test?
    How did I prepare for it?
    What concepts do I still not adequately understand?
After the test, but before the test is handed back:
    How do I feel about the test?
    How well do I feel I did on the test?
    Did I anticipate what was on the test?
    What concepts didn't I know?
    What could I have done differently to prepare for the test?
After the test is handed back:
    Am I surprised at my performance?
    Take two specific errors. For each,
        Reconstruct your thinking for that problem.
        What was behind the error?
        What is your new thinking about the problem?
    What confusion or questions still remain from this unit?

Since the preparation for and taking of exams produced anxiety for many students, yet constituted a large share of their grades, journals provided a place to be intentional about the preparation for and feelings about tests, and then provided the written "product" for later reflection.

The third focused journal activity was a reflective response to the journal as a whole at the middle and end of the semester. The students were asked to read their journal in one sitting, as if they were reading someone else's work, and then react to the reading of the journal. This activity enabled students to assess their own affective and cognitive habits sequentially over the semester and notice changes in them.

In general, the journals were dialogue journals in the sense that they were a two-way conversation between the students and me. Obviously, most of the writing was done by the students, but I collected the journals at random, read them, and responded in a variety of ways: to affirm, to answer questions, to raise questions, or to encourage students to write about a particular process or concept with which they were having difficulty. From past practice, I found that students appreciated that I found their writing worthy to be read, and were eager to see my comments.

AUTOBIOGRAPHICAL WRITING   As the first entries in their journals, I asked students to write three autobiographical narratives about any three mathematical experiences. The narratives were to be told as stories, with as much detail included as possible. I thought this particular form of expressive writing was desirable as the initial activity for several reasons.

These autobiographical narratives gave the students a chance to start with their own experiences and write about something that was familiar to them. Most students were surprised and puzzled when they initially encountered the idea of writing in mathematics class. In fact, many responded that writing belonged in English class and other courses where one explored ideas and wrote essays, but not in mathematics class where the content was numbers. As students captured their feelings about and experiences with mathematics on paper, they realized that the writing process became the vehicle by which they recognized feelings and experiences, and that the written product could become a record for referral and reflexivity.

Writing belonged in English class and other courses where one explored ideas and wrote essays, but not in mathematics class where the content was numbers.

The narratives also provided a source of data about students and their experiences with mathematics, giving me realistic glimpses of such issues as the students' conceptions of mathematics, their attitudes about learning, their perceptions of their own abilities, and their views of mathematics classes and teachers.

FOCUSED IN-CLASS WRITING   On the left-hand pages of the writing book, students wrote entries during class. In contrast to the unstructured writing done in the journals, classroom writing was generally in response to a topic I posed and related to the particular mathematics content currently studied.

I used classroom focused writing in a variety of ways. Sometimes I asked students to write out their versions of a math concept, definition, or process. Or, at the beginning of class, students wrote to find out where they were with respect to the old or new topic. When students ended a class with a short focused writing activity, it afforded them the opportunity to summarize content in their own words, reflect on what they have learned, or raise questions. The focus for in-class writing ranged from more general topics like "What is mathematics like" and "What is calculus," to more specific ones such as "What does it mean for a function to be continuous at $x = a$," "Write out the process of solving $D_x[3x(2x^2 + 3)]$," and "Of what use is the second derivative when graphing?"

As the class and I engaged in expressive writing activities, I realized from both the journal writing itself and from the evaluation of the journal activity by the students at the end of the project that potential benefits existed for the student as writer, for the teacher as reader, and for the classroom due to the reader-writer interaction. (For a detailed description of this empirical study and its qualitative methodology and analysis, see [Rose, 1989].

The writing dialogue established a "correspondence" between each student and me that had multiple benefits for both parties and implications for the classroom rapport and atmosphere. When students wrote, they learned about their feelings about mathematics and the course, the subject matter, their mathematical processes, and their views of mathematics. In addition, what they wrote was beneficial to me, either in responding directly or by taking their concerns into account as I taught the daily lessons.

Benefits to the Student as Writer

As I read my students' writing and what they said about their writing, I was able to categorize ways in which the writing had benefited them. I've labeled these categories affect, subject matter, process, and views of mathematics.

AS STUDENTS WROTE ABOUT AFFECT   Students wrote in their journals almost twice as often concerning their *feelings* about mathematics and/or the course than about the other categories. This is not surprising, since keeping an academic journal was a new experience for most students. When left to do what came naturally, students wrote about how they felt about their previous experiences, what they were feeling about and doing in the course, or about the wide range of emotions they felt concerning various topics in the course. Sometimes they expressed frustration:

> Well, so much for knowing what to do in calculus. I just tried to do the homework and I'm totally lost. I have no idea what I am doing so I gave up. Section 9-7 really confused me. I do not understand composite functions, the chain rule, or the general power rule. I tried to look at the book but that doesn't help either. I am going to have to make an appointment to see you. There is no way on earth that I will pass this test next week. I'm too confused. I can't understand the book so it doesn't help to look at that. I don't know where to turn.

This entry enabled the student to vent his feelings and communicate his frustration. I also took advantage of knowing how the student felt by first legitimizing the feeling and then offering constructive steps to alleviate the confusion. In this case, I urged the student to become specific in pinpointing areas of difficulty so that we could deal with them together. The student made an appointment with me and we worked on the material together before the test.

Other students reported therapeutic benefits in their journals, evaluation questionnaires, and interviews, such as:

> It helps me to get my feelings out more than speaking, cause it's hard for me to talk to people sometimes . . . it brings out what you sometimes keep inward.

> To talk about it [in writing], well it takes the pressure off because you acknowledge that you can't answer it. That you can't do it, and then you can ask whatever you want to ask without feeling . . . stupid.

> . . . helped me calm down. Writing made the course relaxing and easier to bear.

It helps me to get my feelings out more than speaking, cause it's hard for me to talk to people sometimes . . . it brings out what you sometimes keep inward.

As STUDENTS WROTE ABOUT SUBJECT MATTER   Most of the focused in-class writing assignments concerned specific mathematical content being currently covered, so the students were forced to write about specific concepts. Students used their own words to express or acquire understanding about the topic, as illustrated in the following example:

> What is a composite function? What is the chain rule? A composite function is one that contains more than one level of computation, i.e., a quantity raised to a power that is also raised to a power. The chain rule allows us to differentiate a composite function without having to evaluate it algebraically (and in some cases the evaluation is not possible). The chain rule works from the outermost part of the composite to the innermost, finding the derivative of each and multiplying them together.

Not only this student, but others observed that in-class writing provided them with a means to reveal what they understood, to explain what a particular concept meant, and simultaneously to record it for later referral, and to reinforce the learning.

When left to write unstructured spontaneous journal entries outside of class, students in the first weeks of the semester wrote more about how they felt about class and mathematics and about the way they did mathematics than about specific mathematical content. Starting midsemester, however, students increased the frequency and length of entries about specific concepts as the subject matter became more difficult and students needed to increasingly struggle with the material to understand it, as I urged them to become more specific in writing about their problems/difficulties, and as they experienced the power of writing to help them learn mathematics.

Students often wrote about a specific type of problem and identified difficulties they were having:

> I've found increasing difficulty in solving for critical values and I think it's the same old problem—weak algebra. I seem to get stuck every time at the same point factoring $f'(x)$ and $f''(x)$ to solve for x when the equation $= 0$ or undefined. So far the only method for factoring I've been using is the substitution method. Set up the ( ) ( ) and fill them in and multiply to see

if I'm right. There's got to be an easier way. I guess I'll take a look back in this book or Rich's Math Essentials book to see what I can do about it. Once again I've found it impossible to complete the homework (when you spend 1 hour on each problem and still don't solve it—well, it's tough to finish).

In this case, I responded with an example of factoring the derivatives.

Sometimes in the act of writing out a problem, students solved it:

> There's a problem I'm working on right now but I don't have any textbook or homework handy so I'll try to remember it as best as I can. It was an average of a differential $C(x)$ and we had to get the average in terms of x. I know $\underline{C}(x) = C(x)/x$ and $C(x) = 7200 + 60x$. The obvious solution to me seems to be $(7200 + 60x)/x$ but I happened to glance back at the answers and it gave $C(x) = 7200/x + 60$. O.K., I am a dummy. I just figured out what I did wrong. $(7200 + 60x)x$ is the same as $7200/x + 60$, cancel out the x's and walla [he means voila!] See . . . this journal really does come in handy once in a while.

When students reflected on writing about subject matter, they named the following benefits.

Writing promoted understanding. Students realized that when they used their own vocabulary, they constructed a very individualized meaning, like an "inside joke" or a personal translation. They affirmed Vygotsky's (1962) theory that expressive writing represents an expansion of "inner speech;" that is, when students found their own language for mathematical content, they experienced his "deliberate structuring of the web of meaning" and formed bridges between their previous knowledge and new material. Students expressed this in several ways:

> As far as I'm concerned, the textbook is usually gobble-de-gook and I have to put it in my own words before I can understand it.

> Maybe because it was coming from me . . . a lot of times it was in language I didn't understand [the book], so I'd have to translate it into my language so it made it easier that way.

> . . . cause you know what you're talking about. Cause you can understand more in your own words than you can someone else's words. Like when you're writing to a close friend or something, like an inside joke or something, that person understands but if somebody else were to read that, they wouldn't understand that.

Writing facilitated reasoning and problem solving. Students reported that writing helped them break a problem into manageable parts:

> When you look at a word problem, there's no possible way, you can't do that. There's no way you can figure out how something is falling at some rate 10 miles away, and when you write down each step, it's neat to see how everything is intertwined . . . you can break it down into different parts . . . just take one step at a time . . . I sit down and try to do everything in my head before I put it on paper. I do ten steps in my head and then I try to figure out the answer. Then I lose it. When I write about it, I write through each process, you don't skip a bunch of processes.

As students wrote about the process, they benefited because writing forced them to *elaborate* and to be *sequential*. That is, students were forced to slow down their thinking so that they did not lose their thoughts. They wrote step-by-step parts of the problem and named each step. As a result, the icon of the completed process revealed connections among the steps of the problem.

Numerous students named the benefit of "writing to think things through." They expressed it in several forms:

> The times that I get the most out of my journal writing is when I explain processes in my own words and when I try to "dig" myself out of my confusion by writing down the process I'm supposed to use. I need to explain this to myself so I can understand it—I'm glad I have a journal to do that with.

> When writing out what you don't know, sometimes you discover the answer through writing.

> A couple of journals I didn't know what to say. For example, the steps in how to sketch a graph. I just wrote that down in the journal and as I wrote that down, I was actually writing down the steps.

The times that I get the most out of my journal writing is when I explain processes in my own words . . .

In other words, when students wrote expressively, writing enabled them to construct meaning. These comments by students gave evidence of Yinger's (1985) contention that expressive writing provides a continuous flow of new information, called propositional knowledge, which enables linking and organization of information to construct meaning.

Writing was integrative. One student mentioned that writing helped "tie together" the content:

> It's possible to get an A in a course and all you're doing is teaching yourself to get through the test, and two days after the test you don't remember anything. It [writing] helps you to see how it all ties together.

Writing was a revealer of understanding. Students reported that when they wrote about subject matter, both the *process* and *product* of writing revealed what was understood and what problem areas still remained.

> Writing out what I didn't understand helped me see a clearer picture of what I had to do.

> So if I wrote out what I felt and what I knew, and got it right, it helped me know that I could do it and that's when I was getting stronger. And if I wrote down something and it was backwards, then I knew I had to work on that.

> Writing about how I felt about tests, studying, or certain sections and then rereading it helped me to pick out strengths and weaknesses.

Students also mentioned that writing helped them identify areas where they were still having difficulty:

> Try writing out a procedure; if you can't explain it, then it shows what you are having a problem with, and seeing what the problems were helped know what to work on . . . emphasized what I needed to study harder.

In other words, the process and product of writing made evident the gaps or inconsistencies in the student's knowledge. Once this information was revealed, students were in a position to take appropriate steps for a remedy.

Writing was a way to reinforce learning. Students mentioned numerous times the role writing played in reinforcing learning:

> Cause if I just copy it I more or less try to memorize it a certain way so I won't forget it, but if I write it in my own words, I'll remember it like naturally . . . you're remembering your own version of it.

> As you wrote, you had to read what you'd write, so you were actually saying it to yourself over again. It helped.

Students realized that both the process and product of writing facilitated "remembering." As they wrote *in their own words,* the physical act of writing slowed down the processing of the information and better enabled it to be memorized. In addition, the writing as product was read as it was written, thus providing further reinforcement.

Several students even named this benefit as multirepresentational learning:

> It's like, numbers for a problem is one way of solving it and then when you write it in a journal you have numbers and words, like there's two ways of solving it and words are easier to understand than numbers sometimes.

> My mom's a math teacher and she's always said, when you read something you get it that way. But when you read and write, you're getting it two ways and it helps sink in . . . So if read, write, and read it out loud, it's even better, she says.

As students commented on the power of writing to help them remember, they supported Bruner's (1986) and Emig's (1977) contention that writing is multirepresentational—it involves the enactive, iconic, and symbolic modes of learning simultaneously.

> If the most efficacious learning occurs when learning is reinforced, then writing through its inherent reinforcing cycle involving hand, eye, and brain marks a uniquely powerful multirepresentational mode for learning (Emig, 1977, p. 124).

AS STUDENTS WROTE ABOUT PROCESSES   Often students described the way they attempted a particular type of problem:

> . . . One thing that really bothers me is word problems. It is hard to explain why though. I usually read the problem through a couple times and I never know where to start. I then try to define the variables and set up the equations. That is where I have all the trouble because I can usually solve them without any problems. I have trouble trying to sort out all the variables and get them all straight. If I get confused I try reading the problem over again and think about it for a few minutes. I usually get so confused that I give up hope. I can never sort out all the variables and put them into equations. That's my main problem.

For this student, recognizing this area of difficulty was the first step towards dealing with it. I then urged him to pick a specific word problem and attempt to solve it in his journal so that I could help identify the confusion and give suggestions.

At other times, students described their study processes:

> How much do I study? 45 minutes for each class, 1 hour for a test. When? After dinner, before I go to bed. Where? In my room or when I am at work as a door monitor in the LFC. How? I usually try to go through the examples in the book . . .

In this case, I responded back in the journal concerning the inadequacy of his study routine and made constructive suggestions.

As students wrote about the way they studied and learned mathematics, students perceived various benefits for themselves.

Writing promoted concentration. One student described how writing focused attention on learning:

> When you sit down to write a journal entry it helps you to think because you're writing for a purpose ... You block out everything else in the room and if you're concentrating on the one part that you're writing on, you've got your full attention on it, whereas if you're just like doing your homework you might be worrying about the next problem or the past problem.

In other words, as Emig (1977) points out, writing required an integration of the full brain and, in the process, totally engaged the student in the task at hand.

Writing helped retention. Students reported that writing activities helped them "stay with the course":

> ... kept me on the edge when I was on my way to quitting.

> ... helped me stay with the class.

Writing served as a study tool. As students wrote about mathematics, they claimed it functioned in several capacities as a learning tool. For one student, it served as a "pacer":

> Writing journals after each class made it necessary to review new information, ideas, concepts on a consistent time schedule. Journals minimized floundering and kept me up-to-date on concepts we were covering.

Others found the writing helpful as a way to study:

> It helps go over problems one more time and was like a study review ... helpful in preparing for tests.

> Helped study more for tests.

Writing was a "place to do it." Whether it was struggling through a problem or organizing the material, students reported that the journal was a "place to do it." Some mentioned using writing to think their way through problems in previous mathematics courses, but usually on scrap paper that became lost. The journal became a place to organize, synthesize, or reflect, and was then available for later reference.

Writing stimulated the posing of questions. One student mentioned in several data sources that he benefited when he learned to ask questions:

> I started using my writing to learn by asking questions and then by reading the answers you gave me.

Writing as an icon. In addition to the benefits reaped when students engaged in the act of writing, the written record was also named by students as being valuable:

> Helpful for using as a reference ... when I had concepts and step-by-step processes written out, I could look back through for help while doing the homework.

> When I was studying for the test I'd go back and check the processes over. Cause I don't have those recorded in my notebook, which is why I'd like it back before the final.

> They give you a picture of what you're thinking and you can sit back and look at it. You write down what you know, or what you don't know, and instead of just thinking about what you don't know, you have it right there on the paper and you can look at it.

> When we had to write about our tests, like how we felt before and after, then that helped too, cause I could see how did I study for this test, cause if I didn't write that down from one test to the next test, I couldn't really remember. And to see how I studied, then how I felt after the test and was I surprised with my grade when I got the test back, that helped because then I could see, maybe I did this wrong in preparing for this test, so let me try this method now, if that didn't work. So I think that was good.

The students thus reported that they used the written record of their writing in several ways—as a reference in which to look up procedures and notes they had written, as an icon to reflect what had just been written, or as a record of past performance.

WRITING ABOUT THEIR CONCEPTION OF MATHEMATICS   In the unstructured journal entries, students seldom wrote about their views of mathematics. One exception for each student was during the first class when they were asked to freewrite on "Mathematics is like ..." They wrote interesting images, as illustrated by the following:

> Mathematics is like trying to organize and manipulate pebbles on a beach where the sand and stone are the infinite set of numbers and the forces of sea, wind, and movement are the functions and theories of mathematics. Mathematics is as essential as the rhythms of the sea to daily life and work. Mathematics can also be like the sky. Although the sky is always there, sometimes it's clear and crisp—sometimes dark and stormy: the same is true for our understandings of mathematics. Sometimes it takes a bolt of lightning to clear the confusion on a concept or process. Mathematics plays a role in our daily lives just as surely as our environment, whether stormy or clear and bright, our minds and lives tend to mimic nature's cycles.

Sometimes students revealed what they believed was required to do well in mathematics, as in the following entry concerning the importance of "memorizing":

> There is so much material to learn and so much to remember that I can't keep it all straight in my head ... now all I have to do is memorize how to ...

Or students commented on their views about math ability:

> I guess that things like mathematics come natural to some people, but come gradually to others, but never come (what seems like never) to others.

In general, students seldom wrote *about* mathematics. When they did, it was usually as a result of a prompt from me or a topic on the suggestion sheet. This is not surprising in light of the limited exposure students have had to metamathematical topics. The standardized curriculum expects students to *do* mathematics, not to think *about* its nature or raise questions about its existence.

The standardized curriculum expects students to *do* mathematics, not to think *about* its nature or raise questions about its existence.

Although students rarely expressed generalizations about mathematics, they did perceive that writing had an effect on their conception of the discipline:

> Helped me see that math consists of more than numbers, problems, and tests. Helped gain insight on things such as memorizing concepts and reading the whole chapters word for word ... realizing that concepts are just as important in math as well as science and other courses.

What math is and how its purpose is served became less of a burden and more of a curiosity.

Before I thought math was just a bunch of numbers you manipulate to get an answer. After writing about it I see that it's different concepts and strategies. It's sort of like logic and there are a lot of ways to do things.

Writing enlarged students' views of both the content and methodology of mathematics. It enabled students to see mathematics as more than number crunching and single "right" ways to solve problems.

In addition to these advantages, some students recognized unexpected fringe benefits.

Writing improved writing skills. Several students reported that writing had an effect on their writing in general:

Not only did my writing about math improve, but my writing in general. It seemed as though I became more natural at how to write.

Writing promoted independent learning. One student mentioned that writing promoted self learning:

Able to help self instead of having someone else, because I could look back in the journal.

Writing to learn had "transfer" benefits. The writing activities were seen by several students as tools which could be used in other courses:

I've just started doing that [writing out everything for every course] and it's helped a lot.

Writing benefited students personally. Students reported personal benefits from writing. Several mentioned an increase in self-confidence:

I've always had a problem with self-confidence and when I'm writing out my journals I can like see myself progress because everything's in my own words.

I learned more about my strengths and weaknesses.

I think it was an excellent method to help me overcome being shy in class because I wouldn't say anything. If you know your're wrong, you don't want to say something ... I used to get embarrassed at quite a few times because I always say something real fast and I'd be wrong, but in my journals you gave me the impression that if you feel you know the answer, don't be shy, just say it. Just say it, and if you're wrong, then that's the only way you'll learn how to do it the right way. I did that in the journal.

Benefits to the Teacher as Reader

When students wrote, whether it was about Affect, Subject Matter, Processes, View of Mathematics, or the Course, I profited. For everything that the students wrote was feedback to me, and thus could be used for diagnosis and evaluation, increased sensitivity to students and their needs, short-term adjustments in the course, and long-term changes in teaching. Where else could I get this kind of first-hand information from students? In the normal classroom, there is little time to talk to students individually, and the usual written correspondence consists of problems on homework and tests. I benefited from the writing experience in a variety of ways, as illustrated in the following examples.

Everything that the students wrote was feedback to me, and thus could be used for diagnosis and evaluation, increased sensitivity to students and their needs, short-term adjustments in the course, and long-term changes in teaching.

Individual diagnosis and evaluation. When students wrote entries that revealed misconceptions about mathematical content and I read them, I could address those needs, either individually or collectively. The following journal entry gives such an example:

... These values are the critical points. These x values I then sub back into the $f(x)$ to find the point where there is a local max or min. This is done after I make a sign chart to see if and where the function changes signs ... There is also the topic of vertical and horizontal asymptotes. To find the vertical I find out what x value makes $f(x) = 0$ ...

When I read this entry, I realized that this student did not understand the difference between the roles of $f(x)$ and $f'(x)$ in determining critical values. In addition, the student had an incorrect procedure for finding vertical asymptotes. I then wrote back to the student and discussed these misconceptions.

Sensitivity to students/personalization. In an ordinary course, I have little opportunity to come to know each student in a personal way. There are always students who come up after class or visit my office, but many students come to and leave class with little personal contact. When I read their journals, I gathered information about students that enabled me to be more sensitive to them, as illustrated by the following entry:

Today was my first day back since being sick ... I found out that we have an exam on Chapter 9 on Monday. I'm very nervous because I have four other tests that week and I am fighting a losing battle of catch up ...

Upon reading that entry, I initiated contact with the student and postponed his test until later so that he would not be so overwhelmed after his hospitalization with pneumonia.

Another student wrote an entry that sounded like a desperate plea for help:

Class isn't going good, homework is going terrible. I have no time, things are bad—I have got to talk with you or something. I don't know what to do. The next test has me terrified. I don't know what to expect.

Once I read this entry, I wrote him a response that combined encouragement with advice regarding new study strategies. In addition, I made multiple suggestions regarding uses he could make of his journal. He responded positively, and ended the semester quite differently than he started.

I recognized that journals promoted personalization early in the semester. In my field notes I wrote:

... One thing is for sure. I know a lot more about each one now than last Friday. What a difference reading their writing makes. They are all people now and I hope it sets up a good relationship between us. I will be interested to see how much my comments are attended to.

At this point in the semester, the students had only written their autobiographical narratives and a few entries, but already I had gathered personal and academic information about each student. It was like getting new pen pals.

Short-term benefits. When students wrote about their feelings, concerns, and problems, I was often able to make short-term adjustments in the course. Consider the following journal entry:

> . . . I had a hard time keeping up in class when you were going over word problems [related rates] . . . I'm still having a problem with section 10.1 and I think a lot of the people in the class are.

After reading this entry, I became aware of the need to spend additional time on related rates.

In the following journal entry, a student discussed his distress with functions:

> When I get something new to me (something taught to me), I usually pick up on it quickly. But when functions came along I lost it. I just can't comprehend what they are, or what they do, or what their purpose is. I can't determine if an algebraic expression is a function or not without graphing; and then only when I get the graph right . . . [JE# 1,# 11].

After reading this entry, I made the following entry in my field notes:

> When I collected the journals the first time, Greg wrote that he never had really understood functions, particularly when they were involved in word problems. I wrote back that we could talk about them together. The day after I handed back the journals he came up after class and asked when we could get together. We arranged to meet at 7:45 on Wedneseday morning. He came in and we went over basic functions stuff and arranged to meet again at 7:45 on Friday . . .

This example incorporated both the benefits of individual diagnosis and short-term benefits, for I was able to address the student's need plus make some adjustments in the course. When I saw the student's difficulty with functions and realized how important that concept was to the understanding of further topics in calculus, I made time during subsequent class periods to talk about functions. I even used the lab table as a simulated function machine. Without the student's discussion of this in the journal, I would have no way of knowing that this topic needed extra coverage with the entire class.

Long-term benefits. As the students wrote, they revealed insights that were helpful to me as I attempted to learn more about how students thought about and studied mathematics. For instance, I realized the extent to which algebra weaknesses hampered their doing calculus problems, the failure of most students to summarize or synthesize material as they studied for tests, and the appreciation students felt when they were allowed to correct their tests for extra credit. In the following entry, a student described his best study routine:

> My best routine for studying mathematics is repetition. I feel the more I practice certain types of problems the better I become at them. I will go to math labs and even practice problems off old tests to prepare myself for a test. Math is one of the hardest classes for me to study. I feel that in math you either know it or you don't know it . . . [JE# 2,# 3].

As I read the journals, I had the opportunity to collect the views of many students and thus form a composite of the views held by my students.

When I read this entry, I learned this student's conception of what it meant to know and study mathematics—repetition. As I read the journals, I had the opportunity to collect the views of many students and thus form a composite of the views held by my students. This, in turn, influenced how I "talked" about mathematics with students—these particular students and all my students in the future.

Another student gave feedback on one of the tests:

> UGH. This test was terrible. I couldn't believe how tough it was. Plus we just swept over higher order derivatives so I didn't study that much because you said it was easy and thus I thought there wouldn't be hardly any points on the test. But to my dismay there was at least 30 points on the test. What's the deal! If skimming over a part really fast and saying its easy isn't a hint that it's not important, I don't know what is.

This entry caused me to reflect on my treatment of higher-order derivatives and its weight on the test, and to thus consider making changes for the next time I taught the course.

Benefits of the Student-Teacher Interaction

As students wrote in their journals, there were implications for the quality of the classroom. When a student wrote, I read and responded. The student then wrote subsequent entries differently, taking into account my response. The cycle often continued if the student and I chose to let the nature of our interaction partly dictate the development of the dialogue writing. At the same time, I was influenced as I interacted with the student. My increased knowledge of individual students not only changed the way I responded to them separately, but also affected the way I interacted with the class.

BENEFITS TO STUDENTS AS INDIVIDUALS

Better communication. The most frequent interaction benefit reported by students regarded the increased quality of communication between them and me.

> . . . I feel with you I have a personal relationship, and I can talk to you one on one in the journal. In Spanish, I don't feel like that, because after the class is over with, there's no other means of asking her questions without calling her or setting up an appointment with her. But in our calculus class, I have the advantage of using my journal to say, I don't understand this, I don't understand that, could you please help me here? In my Spanish class, I just don't feel that there's a personal relationship like I have with you.

Others cited that journals set up an additional form of communication besides the classroom:

> Cause it sets up a nonverbal communication between a teacher and a student, because not everyone in the class can get involved in the class. There's just too many people. And the teacher verbally in the class can't get to know each student as well as through writing, cause then you can take your time per person and read what they have to say and then respond to them whereas in class you only have 55 minutes to talk to that many people in the class.

As a result of better communication, several additional benefits were made possible.

Better feedback. When the students wrote and I read their entries, my developing acquaintance with them enabled me to give them better feedback. Students noted this benefit on many occasions:

... if you write about the things you don't understand, it helps a lot to get answers back from the teacher, because you can see where you're making your mistakes.

... if we didn't understand something she would write out a solution or give us pointers on how to answer some of our questions.

... many times the reason I wrote what I wrote was that I was hoping to get a response to set me straight.

Open and comfortable relationship between student and teacher. Several students voiced the opinion that writing journals made them feel more comfortable and relaxed with me and the course:

... feel more comfortable with teacher and more willing to voice your differences.

It was as if everyone had met you personally ... we all feel you took a personal interest in us.

Motivation. Another benefit to the interaction named by students involved increased motivation:

When you took the time to answer our questions, that sort of kept me motivated through the course.

Encouragement. The students also reported that when they wrote journals, the interaction between them and me provided encouragement:

Her responses cheered me up. I looked forward to reading her responses. You kept writing back asking me not to give up ... You responded that you were proud of me.

INTERACTION EFFECTS FOR THE CLASSROOM   Besides the benefits to the students individually, as students and I interacted, there were also benefits for the classroom. Some students spoke of the underlying atmosphere:

The more I wrote in my journal, the easier it was to communicate in class. There was a more general understanding.

... affected the classroom atmosphere in that it was easier to communicate because there was more than just a classroom speaking relationship.

The more I wrote in my journal, the easier it was to communicate in class. There was a more general understanding.

Others claimed it helped their attention:

I felt that you knew what was going on between me and math so naturally I paid more attention.

Class participation was also mentioned:

It encouraged me to participate when everyone was asleep and realizing that my input in class was helping—that encouraged more.

Another perceived benefit was in sharing the writing:

In our in-class [writing], you asked somebody to share it. I know that sometimes when A or W, when they, they're smart, and when they read something that they wrote down, I was, like, Oh my goodness. I understood it.

A Writing Illustration

To illustrate how students wrote on a single mathematical topic in multiple categories (in this case affect, subject matter, processes, and feedback on the course) and realized the benefits discussed above, consider the following writing sequence by Pris.

On the first day that related rates were discussed in class, I asked the students to write at the beginning of the period on "What is a related rate problem?" The only exposure they had to these problems was the encouragement to read the textbook before class. Pris showed some understanding of the topic on that first day while writing an in-class entry:

A related rate problem is one for which the solution is found by taking the derivative (instantaneous rate of change) and to do so we have to set up an equation showing the relationship of a given rate to the data provided for us.

Here she used the focused writing to get down on paper what she knew about related rates, and consequently she revealed to both herself and me the present state of her understanding.

In her subsequent unstructured journal entry, she became more explicit about the procedure for solving related rate problems:

What is the procedure for solving a related rate problem?
1—read the problem
2—read the problem again
3—write down all the information given
   a) create a diagram of what is happening
   b) fill in as much data as possible
4—write a "find" statement
5—create an equation with what is given to you and what you need to find based on what is known of the diagram
6—attempt to take the derivative and solve
7—if the results of 6 don't make sense, go back to 5

Here Pris expanded upon what she wrote in her in-class writing by creating a procedure for herself. This was her own, for the textbook did not list one.

Pris' next unstructured journal entry showed a mixture of content, process, and affect on the same topic:

I am becoming incredibly frustrated with my homework. I'm making some kind of technical error in almost every problem I attempt. I get the picture, "find" statement, and variables all listed nicely. But either I'm using the wrong equations (not likely) or I'm solving them incorrectly (more likely). I'm afraid I don't know where to start looking to find my error. Tomorrow's class will help (but that doesn't help my homework tonight, though). Maybe I'm just confusing myself—making things tougher for myself than they really are.

The next class produced the second focused writing about related rates. Notice the increased details and personalization of the process compared to the previous description of the procedure:

When I try to do a related rate problem: First, I read it over and then I read it again. Then I try to draw a diagram of what's happening—and I usually get an accurate diagram down on paper. Next I "fill in the blanks"—all the information bits and pieces relating to the problem are put into the diagram. Again, I can handle this pretty well. However, now comes the tough part. I know that the equation should be something like a volume, area, or Pythagorean theorem, so I write down a "logical" choice for the equation and I get stuck. Do I differentiate now or do I substitute in first? And if I substitute in first I get a 0 for those numbers. If I don't, I get too many variables to solve for what I need. Frustration is working on one of these problems for a long time and then not knowing where to go next.

At this point, Pris moved past the more technical description she gave previously and grappled with the "nitty gritty" of the related rates process. By writing this out, however, she had the opportunity to reveal her questions and frustrations to both herself and to me. In this case, it reminded me that the notion of when to substitute the given values is a confusing one for students and should be dealt with during class for everyone's benefit.

During the class period in which Pris wrote the previous focused writing entry at the beginning of class, more related rate problems were worked on together. That night for her next unstructured journal entry, Pris wrote again on related rates:

What we did to solve # 7 today:

> After reading the problem, the information was used to create 2 diagrams—one of the motion with variables listed and one with the specific information labeled on a "stopped" picture. Then we wrote a "find" statement. Next, we used the Pythagorean theorem as our equation and differentiated it in terms of time. Then we substituted in our "stopped" information and solved for what we decided we needed in the find statement (db/dt). I wish it were that easy when I'm struggling with them by myself.

It is interesting to note that Pris got even more specific in this description; in addition, she appeared to have answered the question she raised concerning substitution.

Ten days later, Pris wrote another journal entry about related rates, reporting a new awareness about the process and her feelings concerning writing about it:

> I have figured out what I was confusing in related rates problems: When I was taking the derivative, I was confusing the difference between a constant's derivative as opposed to a constant times a variable's derivative, i.e., $1/3(\pi)r^3$. I was taking the derivative of (π) as 0—which was scrambling my results. It does help me to write down the different processes and definitions; but sometimes I can write down a process (like for related rates) and still not quite be able to use it to solve a problem. Definitions come easily, though, especially in this course, which has taken us in sequence so that we understand limits before derivatives and derivatives before differentials.

Here Pris reflected upon her confusion with derivatives and constants, and appeared to feel better about her ability to now carry out the related rates process. Her assessment about writing identified both the positive role it was having in her learning, and that it does not necessarily guarantee automatic understanding without a struggle for meaning.

This sequence of entries, both focused in-class and unstructured journal entries, are typical of Pris' use of multiple categories when she wrote. For Pris as the writer and for me as the reader, there was a story unfolding about related rates. The process was stated and restated several times, each time dealing more specifically and personally with the hard issues of the procedure. Both the cognitive and affective components of related rates were revealed and Pris' struggle was indeed felt.

Conclusion

Using writing to support the learning of mathematics has great potential as an educational tool—for the student as writer, for the teacher as reader, and for the classroom due to the writer-reader interaction. When students write, they can be encouraged to express and reflect upon their feelings, subject matter, processes, and views of mathematics, and consequently cope with negative emotions, learn new content and skills, ask questions, and reconceive their beliefs about mathematics. As teachers read the writing, they are exposed to individual needs, common difficulties, and feedback on the course; with this information, teachers can become more responsive to short-term adjustments in the course and long-term improvements in teaching. As students and teacher engage in a dialogue, the interaction can produce a more personal, cooperative, and active learning environment.

Like any other tool, however, there are numerous factors which determine its effectiveness. Teachers are often concerned that incorporating writing in the classroom will take precious time from "covering" the content and require extra time for reading students' writing. Granted, writing assignments will require small adjustments in the manner in which teachers use their time, both in and out of the classroom, but the rewards are well worth the effort as both the affective and cognitive needs of students are realized and met.

Granted, writing assignments will require small adjustments in the manner in which teachers use their time, both in and out of the classroom, but the rewards are well worth the effort as both the affective and cognitive needs of students are realized and met.

Although teachers can provide expressive writing activities and can facilitate growth and change in students' writing by establishing a caring and trusting environment, students will respond to the opportunities in different ways. Variables such as personality, learning and teaching styles, mathematical background, and gender may affect the impact of writing on learning, yet students report positive benefits as they write in their words about their needs. As unique individuals, students must learn by experience and "grow into" writing. External circumstances may encourage and foster it, but writing needs time. It is a nonlinear, dynamic, mysterious, and developmental process.

Writing activities in mathematics instruction are adaptable to the needs of both teachers and students. Regardless of the level or content area, writing works because whenever students write, they individualize the instruction and construct personal meaning with *their* own language. That is what learning is all about.

BIBLIOGRAPHY

This bibliography contains not only references from this paper, but other citations regarding writing and mathematics.

Abel, Jean P., "Using Writing to Teach Mathematics," paper presented at the Annual Meeting of the National Council of Teachers of Mathematics, Anaheim, California 1987.

Bell, Elizabeth S. and Ronald N. Bell, "Writing and Mathematical Problem Solving: Arguments in Favor of Synthesis," *School Science and Mathematics* 85 (March 1985): 210–221.

Borasi, Raffaella and Barbara J. Rose, "Journal Writing and Mathematics Instruction," *Educational Studies in Mathematics* 20 (November 1989): 347–365.

Britton, James B., Tony Burgess, Nancy Martin, Alex McLeod, and Harold Rosen, *The Development of Writing Abilities (11–18)*, Macmillan Education Ltd., London, 1975.

Bruner, Jerome S., *Actual Minds, Possible Worlds*, Harvard University Press, Cambridge, Massachusetts, 1986.

Burton, Grace M., "Writing as a Way of Knowing in a Mathematics Education Class," *Arithmetic Teacher* 33 (December 1985): 40–45.

Countryman, Joan, "Writing to Learn Mathematics: Some Examples and Strategies," paper presented at the Annual Meeting of the National Council of Teachers of Mathematics, Anaheim, California, 1987.

Emig, Janet, "Writing as a Mode of Learning," *College Composition and Communication* 28 (May 1977): 122–127.

Evans, Christine Sobray, "Writing to Learn in Math," *Language Arts* 61 (December 1984): 828–835.

Fulwiler, Toby, "The Personal Connection: Journal Writing Across the Curriculum," in T. Fulwiler and A. Young (eds.), *Language Connection: Writing and Reading Across the Curriculum*, National Council of Teachers of English, Urbana, Illinois, 1982: 15–31.

Geeslin, William E., "Using Writing About Mathematics as a Teaching Technique," *The Mathematics Teacher* 70 (February 1977): 112–115.

Goldberg, Dorothy, "Integrating Writing into the Mathematics Curriculum," *Two-Year College Mathematics Journal* 14 (November 1983): 421–424.

Hirsch, Lewis R. and Barbara King, "The Relative Effectiveness of Writing Assignments in an Elementary Algebra Course for College Students," paper presented at the Annual Meeting of the American Educational Research Association, Montreal, Quebec, Canada, April, 1983.

Johnson, Marvin L., "Writing in Mathematics Classes: A Valuable Tool for Learning," *Mathematics Teacher* 76 (February 1983): 117–119.

Kennedy, Bill, "Writing Letters to Learn Math," *Learning* 13 (February 1985): 58–61.

Kenyon, Russell W., "Writing in the Mathematics Classroom," *New England Mathematics Journal* (May 1987): 3–19.

King, Barbara, "Using Writing in the Mathematics Class: Theory and Practice," in C.W. Griffin (ed.), *New Directions for Teaching and Learning: Teaching Writing in All Disciplines*, Jossey-Bass, San Francisco, 1982: 39–44.

Lipman, Myra R., "Mathematics Term Paper!" *Mathematics Teacher* 74 (September 1981): 453–454.

McMillen, Liz, "Science and Math Professors Are Assigning Writing Drills to Focus Students' Thinking," *The Chronicle of Higher Education* (22 January 1986): 19–21.

Mett, Coreen L., "Writing as a Learning Device in Calculus," *Mathematics Teacher* 80 (October 1987): 534–537.

Montague, Harriet, "Let Your Students Write a Book," *Mathematics Teacher* 66 (October 1973): 548–550.

Nahrgang, Cynthia and Bruce T. Petersen, "Using Writing to Learn," *Mathematics Teacher* (September 1986): 461–465.

Pallmann, Margot, "Verbal Language Processes in Support of Learning Mathematics," *Mathematics in College* (Fall 1982): 49–55.

Powell, Arthur B., "Working with 'Underprepared' Mathematics Students," in Mark Driscoll and Jere Confrey (eds.), *Teaching Mathematics*, Heinemann, Portsmouth, New Hampshire, 1986.

Rose, Barbara J., "Using Expressive Writing to Support the Learning of Mathematics," PhD dissertation, University of Rochester, 1989.

——, "Writing and Mathematics: Theory and Practice," in *The Role of Writing to Teach Science and Mathematics*, Teachers College Press, New York, 1989.

Sachs, Marvin C., "Writing in the Mathematics Curriculum: Back to Basics or Something Different?" paper presented at the Ninth Conference on Curriculum Theory and Practice, Dayton, Ohio, October 29, 1987.

Schmidt, Don, "Writing in Math Class," in Anne R. Gere (ed.), *Roots in the Sawdust*, National Council of Teachers of English, Urbana, Illinois,: 104–116.

Selfe, Cynthia L, Bruce T. Petersen, and Cynthia L. Nahrgang, "Journal Writing in Mathematics," in Art Young and Toby Fulwiler (eds.), *Writing Across the Disciplines*, Boynton/Cook Publishers, Inc., Upper Montclair, New Jersey, 1986: 192–207.

Shaw, Joan G., "Mathematics Students Have a Right to Write," *Arithmetic Teacher* 30 (May 1983): 16–18.

Stempien, Margaret and Raffaella Borasi, "Students' Writing in Mathematics: Some Ideas and Experiences," *For the Learning of Mathematics* 5 (November 1985): 14–17.

Trivieri, Lawrence, "Writing Across the Algebra Curriculum," paper presented at the Annual Meeting of the New York College Learning Skills Association, Ellenville, New York, 1986.

Vukovich, Diane, "Ideas in Practice: Integrating Math and Writing Through the Math Journal," *Journal of Developmental Education* 9 (January 1985): 19–20.

Vygotsky, L. S., *Thought and Language*, MIT Press, Cambridge, Massachusetts, 1962.

Watson, Margaret, "Writing Has a Place in a Mathematics Class," *Mathematics Teacher* (October 1980): 518–520.

Yinger, Robert J. and Christopher M. Clark, *Reflective Journal Writing: Theory and Practice*, (ERIC Document Reproduction Service ED 208 411, 1981).

Rewriting Our Stories of Mathematics

Linda Brandau
The University of Calgary, Calgary, Alberta, Canada

Introduction

We all have a story to tell of our mathematical lives, a story comprised of our personal school experiences, everyday experiences, and professional experiences. We all have also created a story of mathematics, our own perception of what mathematics is, our own understanding of mathematical concepts. The kind of mathematical life story we have to tell and the kind of story of mathematics we have created work together to influence what we think of ourselves and the kind of career we choose.

This essay explores these two kinds of stories in a specific context—that of university students writing about their memories of mathematical experiences while working on and writing about mathematics problems. As a result of this writing, students' understanding of mathematics has been enriched. Even more important has been the transformation of their mathematical life story and their story of mathematics. But I am getting too far ahead of my story.

The kind of mathematical life story we have to tell and the kind of story of mathematics we have created work together to influence what we think of ourselves and the kind of career we choose.

Since 1984 I have been a teacher educator who works with elementary school teachers and with students studying to be elementary teachers. Previously I taught grade seven and eight mathematics, a career which grew out of having a BS in mathematics. The roots of my work with elementary school mathematics lie in my work with junior high school students, ones who would respond (to my great disturbance) to my efforts to help them think for themselves by saying, "You're the teacher, tell us what steps to do." My disturbance with this attitude did not go away, so when I worked on my PhD in education, I focused on tracing the beginnings of this phenomenon. While doing so, I learned that I needed to be concerned about children in the early years of elementary school.

At the same time, I was observing the development of my son's views about mathematics throughout his years in elementary school. Even in grade three, he had no interest in any other kind of understanding of mathematics than learning the mechanics needed to complete the blanks in his workbook. He would respond to my efforts at evoking a deeper level of understanding with exasperation, "Mom, just tell me the steps to do so I can get this page done!"

What has amazed me over the last 20 years is the similarity of working with students of any age from 6 years old to 16 years old to 46 years old. What I have observed in too many of these students is a superficial understanding, a view that mathematics means "memorize some rules and procedures."

When I listen to colleagues, whether they are mathematics educators, classroom teachers, or mathematicians, I hear variations of my observation over and over and over. As a result, I have been led to the following concerns: What is occurring? What is happening in classrooms and in the culture? Even with all of the so-called educational "reforms," why am I still hearing these same stories? Why are all of us in mathematics and mathematics education still telling these same stories?

In an effort to deal with these concerns, I decided to work with elementary teachers and student teachers at the level of their beliefs about mathematics. My reason for choosing elementary teachers seemed obvious. This was where schoolchildren were beginning to memorize rules and procedures. My reasons for working at the level of the teachers' beliefs also seemed obvious. If teachers believe mathematics consists of memorization of rules and procedures, then their teaching and lessons will reflect this belief; it is likely their students will also have this belief. If teachers can become more consciously aware of their beliefs and can begin to work on transforming them, then their teaching will reflect this awareness and work.

Strongly interconnected with beliefs is a teacher's understanding of the mathematical concepts he will be teaching. Recently I asked a group of elementary teachers what division meant. Most of the responses were similar to: "In a problem like 556 divided by 7 you see how many times the 7 goes into 55, which is 7 and then . . ." I stopped the students at this point and asked the question again. This was met with silence. The students assumed they had already answered my question. What was fascinating to me was the form their answers took, that of the steps to the long division procedure, but nothing was said about the concept of division, that is, the sharing or portioning of a quantity into equal sized groups. (I challenge readers who teach any level of mathematics, even university, to ask their students a similar question, such as the meaning of differentiation or integration in calculus. My intuition, based on listening to university mathematicians, would be that students would also respond by describing only steps to the procedure.)

Even more fascinating was what happened when we continued our discussion using the problem 28 divided by 3. One of the students took 28 small blocks, divided them into nine groups with three in each group. When I said that her physical representation did not match the problem given, she argued that she could see from her groups that nine with one left over was the answer and that it didn't matter whether she had formed three groups with nine in each or nine groups with three in each. I then contextualized the problem as $28 divided among three people versus $28 divided among nine people. She agreed that now it mattered if there were nine or three groups (in terms of the amount of money each person gets) but still insisted that it did not matter what representation was used when she did the problem 28 divided by 3.

If teachers believe mathematics consists of memorization of rules and procedures, then their teaching and lessons will reflect this belief . . .

I share this experience because it illustrates the kind of story of mathematics that many students have created, a story that defines mathematics as symbols disconnected from physical world experience. Although it may seem trivial, on the surface, that $28 divided among three people differs from the abstract problem 28 divided by 3, the differences become important in elementary school teaching. When teachers, and students, view abstract numerical symbols as inherently different from real world uses of these symbols, then it is too easy for them to create an incoherent and meaningless story of mathematics. (This is not an isolated story either. Many students

The Mathematical Past in the Present

The method I use to work with students at the level of their beliefs is to create a situation where they are propelled into their mathematical life story and into their story of mathematics. This situation is journal writing, where students write not only about their thoughts and feelings about mathematics, but also are required to work problems in mathematics. The problems used are ones in Mason, Burton, and Stacey's *Thinking Mathematically*, a book that explores the reader's feelings and emotions about solving problems. The problems in this book are more complex than what students are used to solving. They sometimes require hours of work to solve them, and hence can be seen as a response to criticism from educators, such as Schoenfeld [6], who accurately portrays the student's belief that mathematics problems are always solved in less than 10 minutes, if they are solved at all.

Students work in their journal every week, outside of class, for an hour or two and turn it in to me for my reactions, feedback, and comments. Students choose the problems on which they work, although I suggest that they begin with the first chapter in the book. An important part of making this choice is that students must decide for themselves when to give up trying to solve a problem, or when to leave a problem for a while and then come back to it.

A final, important aspect to the book, *Thinking Mathematically*, and its problems is that answers are often not given. This feature has caused much debate among the students and myself, both in the journals and in class. The discussion centers around how answer-oriented their mathematics classes have been and also around how frustrated they feel having worked for hours and then not knowing if an answer is right or not. For many of the students, this frustration does not lessen at all throughout the entire semester. But some students cope well and even come to the point of not needing to look for an external answer from a key; they come to trust their own sense of satisfaction with what they have obtained.

It is important to point out that this variation in student reaction is reflected in the variety of journals written. I make a special effort to emphasize individuality and that I do not think there is "one right way" to participate in the journal process. Even so, I always have many students ask if they are doing their journal properly. I see this question as integrally related to the story they have created, one that says that mathematical problems have one right answer (and one right way of solving them). Therefore, if I am to help students transform this story, it is crucial for me to be accepting of any and all journal writing, no matter how sparse or minimal.

Because of the enormous variety in these journals, it would be difficult to provide indicative "samples" here. But even more importantly, a smattering from a variety of journals will not provide an adequate picture of the process, of the themes elicited and how they interact to transform students' stories. Hence, in the rest of this essay, I will focus on the work of one student, Allen. The problems I have chosen to share are not the only ones he solved, but were chosen to make some specific points in this essay. To give some idea of the development of Allen's stories (at the level of both life experiences and mathematical understanding), I present them in chronological order.

First, there is one of the problems labelled "hidden assumptions": Nine dots in a three-by-three square array are to be joined by four consecutive straight lines, without removing the pencil from the paper.

At first, I only had the nine dots and I was getting terribly frustrated in being unable to "get" the answer. I suppose that I was assuming that I was not allowed to go beyond the limit of the nine dots. Once the other dots were added it made it easier for me to solve this problem. I recall teachers using these types of problems in elementary school—they were sort of "trick" questions—where the teacher posed the problem and, at least it seemed to me that they (the teachers) took great delight in holding the answers while we poor students struggled—very often unsuccessfully—to "get" the answer. It was like, "You poor students here is the answer—aren't I wonderful?" type of attitude put forward by these teachers. None of them ever talked about assumptions, etc.

The evocation of these memories is crucial to the present solving of mathematics problems. Here we see Allen recalling this type of problem as a "trick"; yet at the same time, he is beginning to question this past view, one which may have blocked him from solving such problems because it was only the teacher who held the answer to such problems. Now he begins to see these problems as having the purpose of questioning one's assumptions.

Another evocation of a past personal experience occurred for Allen while solving the "paper band" problem: take a thin strip of paper, about 11 inches by 1 inch (28 cm. by 2.5 cm.) and fold it as shown. Join the ends to make a band. Seek some questions.

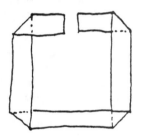

How interesting it is to fold paper in the late afternoon—maybe anytime.

I was wondering if there was some way of predicting how large the space between the ends of the paper band would be—or overlap. I recall doing things like this in school many years ago and I'm sure there were ways of compensating for errors during the folding process. Now why do I call deviations "errors"? I would suspect that the width of the paper would have a bearing on the outcome as much as the total length of the band.

Where one starts folding! Why not start in the middle and fold symmetrically. Then, at least, you would end up (sorry) with the folded ends being exactly the same length. This sort of observational stuff is important to me. This exercise reminds me so much of when I used to make water bombs out of paper when I was a kid. I've enclosed one here for demonstration only. I used to spend hours folding these things—attempting to make bigger—(bigger in those days was better in terms of water bombs)—ones. Paper, shape, folds, volume were variables—of course then I didn't call them variables—that I liked; not x, y, and z. All in an effort to improve the paper water bomb.

When I made these toys I was always trying to improve them. I did not know what I wanted in specific terms but I realized also that I would recognize it when I saw it. I see now that in those days I was actually asking a lot of questions about water bombs—the general one being "How does one make a better water bomb?" "What would it look like?" "How do you fold the paper to get the best water bomb?", etc. I don't know if I ever got a "better" one but I realize that I felt good about going about making attempts in this direction.

One development we see here, perhaps an influence from the previously quoted problem, is Allen's questioning assumptions. That is, he gets the idea of beginning in the middle and folding the paper symmetrically. He also relates this current experience to his memory of making water bombs out of paper. Since the making of water bombs was meaningful for him, this problem also becomes meaningful. He is bringing what worked in the past to the solving of a present problem. In this experience, there is also an awareness that he was doing mathematics in the past (the idea of variables) but did not know it until now. This awareness helps Allen to define the past and to redefine his present perception of mathematics.

Another theme here involves the development of internal trust, that intuitive sense that leads us to knowing when we have an acceptable solution to a problem. When Allen speaks of his continual work, as a child, trying to improve his water bombs, even though he didn't specifically know what he was looking for, he did know that he "would recognize it when I saw it."

This theme of internal trust has become a crucial one in this journal writin; as mentioned previously, *Thinking Mathematically* does not always offer answers to problems posed. Students learn that they need to decide when they are satisfied with an answer (or check with one another) and that they need to learn how to make this decision.

In the next problem Allen solves, the theme of internal trust recurs.

I don't feel really good about this "answer." I feel I somehow just stumbled onto it. I thought that perhaps a rule could be arrived at by a systematic approach! My method was more trial-and-error which I suppose is a system. Maybe I want to be able to "get" answers in a more elegant fashion. Also, finding this answer does not leave me with any sense of pleasure or confidence or satisfaction. To me the feeling I'm left with is not unlike that which I used to get when working math in high school.

The evocation of feelings similar to what Allen felt in high school is important, as seen in his work on the very next problem, "Black Friday": A Friday the thirteenth is known as black Friday. What is the most/least number of black Fridays you can get in one year? In a twelve-month period?

The reason that some Fridays are "Black Fridays" and others are not is because some months have 30 days, others have 31, etc. Maybe by looking at the number of days in each month we can get a clue as to the number of "Black Fridays" there are in a year.

(At this point he lists the months and the number of days in each.)

So what! I sense that "I'm getting stuck" feeling coming on. It erodes my confidence which makes me want to give in and be comfortably "stuck."

But if I just say that Jan 1 is a Friday, what then? Well, Jan 1 = friday, + 14 days; Jan 15 = friday, + 14 days; Jan 29 = friday, + 14 days; Feb 12 = friday, + 14 days; Feb 26 = friday, + 14 days.

This is one way to get an answer. However there must be a better way! It just occurred to me that there are 26 "14 day" periods in a year. Is this important? I don't know. Still stuck. (I'm frustrated and angry!!)

(He continues the next week after I respond by asking about his anger.)

I'm angry because I'm not "getting" the answer. If nothing else my math education—no—let's call it training—my math training has drilled into me that unless you solve the question (i.e. get the right answer) then you have failed. Also, I got angry because the realization that I had not solved the problem brought back to me the exact feelings of frustration about math that I used to get so many years ago when I was in elementary and high school. So really it is not anger directed at a wrong answer but perhaps disappointment that after all these years I have not gotten "good at math." Angry too because, for just a brief moment, I was reminded of how dumb I used to feel when I did not do well in math—get the right (correct) answer. My past seems to have a terribly strong grip on me yet!

Allen's anger and frustration, an intermingling of both past and present feelings, is a necessity in the transformation of both his mathematical life story and his story of mathematics. While solving the problem, he switches back and forth between his frustration with the problem and his solution of it. Although he has trouble with the problem, he still continues to work on it. He is beginning to learn that he can do mathematics even when it becomes frustrating.

At the same time, there is an evocation of memories of frustrating feelings from elementary and high school, and then memories of how "dumb" he used to feel. Because these feelings are being evoked while he is currently solving a mathematics problem, they can be examined for what they are—past experiences and memories. It is important for him to retell, to relive this past, and while doing so to begin to disentangle the memories from the subject matter of mathematics, to become aware that these memories are of himself, not of the mathematics. Often students' memories of math classrooms get translated into the story line: Math is awful, horrible, too difficult, and so on. These stories have evolved from student's feelings about themselves in the situation, not about the subject of mathematics.

Writing to Understand Mathematics

The journal process described above has enriched Allen's understanding of mathematics in a way very different from other forms of teaching and learning. This process personalized the mathematics. It required him to get actively involved, engaged, often lost in the problems he was trying to solve. If we view "understanding" as "grasping the meaning of," Allen has been involved in making his own meaning of what he was doing. He discusses this idea in his final journal entry, a reflection about his work throughout the semester.

At this stage in the journal I would like to think that I now approach math problems differently. I use the word "approach" very carefully and consciously here. Prior to reading Mason's book I used to think that math problems were somehow written in stone in terms of interpretation. *Thinking Mathematically* made me think about this approach. That is, I was forced to consider that problems may be interpreted more than one way. And what are the implications of this for the student (like me) who has been taught math in a very traditional manner? Well, at first glance it would suggest that this liberal attitude (or approach) is a good thing. That is, the student is freed from trying to figure out what the question is "after." The student is not being dictated to by the question—in reality the teacher—anymore. He/she is now free to put his/her own meaning into the questions and from there solve it. In other words, solving math problems changes from a one-way process (question →student) to a two way process (question ↔ student). If I may, it democratizes mathematics for the student.

However, there is a catch! It is that in the more general world where there is freedom and democracy, there are also responsibilities and costs. What, one might ask, are these new costs? They are that the student is now required to put more of himself/herself into the question-answer process. No longer can the student sit back and say "I can't get it! I just don't know what they want." By approaching math in this new fashion (i.e. personalizing or democratizing it) students must take a more active role. And it is a role that many students may be reluctant to play because it requires risk taking, and the taking of risks, although offering the possibility of benefit, is potentially costly as well. That is, the student's input to the math question—his/her interpretation—reveals a personal side of that person; a side that he/she may not want exposed, especially if he/she feels that he/she did not somehow answer the question "right."

I know that several times (at least) as I was doing the problems in *Thinking Mathematically*, I would come across a certain question and I would avoid it by simply going on to the next question. What was going on in my mind when this would happen? It seems to me that I would view the question (problem) and right away make a snap decision as to whether or not I could answer it (i.e. get the "right" answer). In other words, for me in the early problems of the book, I was saying that math problems were only worth attempting if there was a reasonably good chance that I would get the right answer. The problems that I skipped in the book were usually passed over for that reason. Towards the end of the book, however, I think that I was willing to try problems even if I didn't like the looks of them initially. Sometimes, as in the case of the "Black Friday" problem, I became very frustrated and somehow felt that all my efforts were a waste of time. Again, I found myself thinking in the old ways. Why? Well, maybe it was because I was only reading the question and not allowing for my unique interpretation to take place. (↔)

The "Furniture" question is interesting in that the solution was that there may not be a solution. My training in math (in elementary and high school) was that there is always a solution to math problems. This training got in the way of me being able to say, finally, that "I don't believe it can be done." Mason's book, perhaps, has taken me forward a little in stepping backward. By this I mean I am able to look at problems now the way (maybe) I should have been taught when I was a kid. Then problems with no answers were, of course, "trick" questions. Excuse the term, but *Thinking Mathematically* has deprogrammed me in terms of the way I have been trained to solve problems. For example, I don't recall teachers putting much emphasis on diagrams then and manipulatives were definitely not encouraged in any math I did at that time. For this reason "booklets," "paper bands," and "multifacets" were quite enjoyable for me. Enjoyment—fun—satisfaction. These are the things that *Thinking Mathematically* has helped me experience in math finally.

As Allen indicates, by being required to put himself into each problem, he could no longer distance himself from mathematics. No longer was he searching for the answer given in the textbook or by the teacher. He was now creating his own interpretation, his own sense of what appropriate answers would be. He also learned that some problems do not have an answer at all.

This change of focus for Allen illustrates an enriched understanding of mathematics, one that has moved away from solely an answer focus to one that focuses on exploring ways of solving a problem, with exploration being the key here. Because he began to feel free from searching for "the" answer, he could immerse himself in the problem, spend time with the mathematics, not with the search for answers.

Jean Piaget [5], Jerome Bruner [1, 2], and Seymour Papert [4], for example, have taught us that our most enriched understanding occurs when we are actively involved in our own learning. Although a good lecture-discussion class format can actively engage us, it is a different kind of engagement. When a student watches the teacher solve a mathematics problem, and can follow the teacher's method, a certain kind of understanding occurs. It is an understanding that relates to being able to follow another person's thinking process.

But when students struggle with problems on their own, another kind of understanding is required, one that involves being able to create one's own process. This is why we often hear students say, "I could understand how to do the problems when you demonstrated them, but I have no idea of how to do them on my own." On our own, we are required to lead ourselves, to create our own path to a solution. When we listen to others, we follow, we are led on their paths.

Writing is central to the understanding process in general. We often do not know what we think until we need to find the words to put to paper. Then seeing the words on paper gives meaning to what may have been obscure before. This interactive process—of searching for words, writing them, reading them, altering the words, and so on—enhances our understanding of the topic about which we are writing.

In the solving of mathematics problems in particular, writing can also enhance understanding. Although we can be actively engaged in mathematical activities which do not necessarily involve the kind of writing described in this essay, the kind of process I use seems to be unique. What occurs is the capturing of thoughts on paper while one is having the thoughts and while one is in the process of solving a mathematics problem. In trying to find the right words, students grapple with the meaning of the problem they are solving, and hence are grappling with their understanding.

Writing is central to the understanding process in general. We often do not know what we think until we need to find the words to put to paper.

It is important to reemphasize briefly that the tasks or problems given to students are also central to their understanding. They need to be problems that are complex, that evoke struggle, require decision making, and take time to solve. If students continue to see problems that can be solved in only ten minutes or less, then their stories of mathematics stand little chance of being transformed.

Transforming Stories

At the time of the writing of this essay, it is a little more than a year since Allen completed the course where we used the journal writing. He is currently writing more about his past mathematical and other educational experiences, and is continuing to deal with his feelings about himself.

He says that what he remembers most about the work in *Thinking Mathematics* is learning that his own assumptions and interpretations of problems had validity. Because he was "not searching for someone else's answers," he gained confidence in himself, in his own answers.

His story of mathematics and his mathematical life story are in the process of being transformed. He says that he always had a romanticized version of what mathematically smart people could do. That is, he thought they did not do homework, never struggled with problems, knew mathematics instantly. His work and the views proposed by Mason, Burton, and Stacey [3] have dissembled and transformed his romantic story. He now sees that it is legitimate to "be stuck," that even mathematicians get stuck when they work. This is a rewriting of his story of mathematics, one resulting from *Thinking Mathematically*'s emphasis on what to do when one gets stuck.

We transform, rewrite these stories by reexperiencing the past while presently solving mathematical problems. The current problems evoke the same struggles, frustrations, and anger that we felt in the past.

He tells me that the journal writing has helped him gain confidence in his abilities, but that he does not feel "completely healed" yet, that he still does not believe that he "is good in mathematics." This observation, one that relates to the rewriting of both mathematical life stories and stories of mathematics, can be related to the image of an echo as follows. We transform, rewrite these stories by reexperiencing the past while presently solving mathematical problems. The current problems evoke the same struggles, frustrations, and anger that we felt in the past. We see that we can be angry and frustrated and still continue to solve a problem. The stories begin to be transformed. Yet like an echo, every time we try (in the present) to solve certain problems, we reexperience the past. This reexperiencing occurs over and over like an echo does. Yet, like an echo, the loudness of the anger, struggles, and frustrations lessens each time, until finally they are heard no more.

However, for some students, the echo remains. Not all students have the same experience as Allen in transforming their stories. When I propel the students into this process, I try to make it clear that they will rewrite and recreate their views in their own way, in their own time. As a result, many students felt pushed too far, too fast, too hard. They feel like they have been thrown off a cliff and are flailing around trying to cope with this way of learning mathematics. Even though I am there to support them so they will not fall into some dark abyss, many are not ready for this kind of learning experience. It is my place to support whatever kind of work they do in the journal. Many get blocked in the rewriting of their stories of mathematics, but they need to feel that that is all right. They need to feel that I am creating a situation for them to rewrite their own stories and that rewriting will take whatever form is appropriate for them at the moment. At least the opportunity is offered. At least there is the chance for a rewriting.

REFERENCES

1. J. Bruner, *On Knowing: Essays for the Left Hand*, Harvard University Press, Cambridge, Massachusetts, 1979.

2. J. Bruner, *Actual Minds, Possible Worlds*, Harvard University Press, Cambridge, Massachusetts, 1986.

3. J. Mason, L. Burton, and K. Stacey, *Thinking Mathematically*, Addison-Wesley, Reading, Massachusetts, 1985.

4. S. Papert, *Mindstorms*, Basic Books, New York, 1980.

5. J. Piaget, *To Understand Is to Invent*, Penguin Books, New York, 1976.

6. A. Schoenfeld, "Metacognitive and Epistemological Issues in Mathematics Understanding," in E. Silver (ed.), *Teaching and Learning Mathematical Problem Solving: Multiple Research Perspectives*, Lawrence Erlbaum Publishers, Hillsdale, New Jersey, 1985.

Writing in Mathematics: A Vehicle for Development and Empowerment

Dorothy D. Buerk
Ithaca College, Ithaca, New York

For me mathematics is most like:

> an assembly line where a large group of people perform the exactly same task day after day, year after year.

> a giant classroom with millions of men reciting the Pythagorean theorem.

> the military draft: the amount of interest I have for math is comparable to how much soldiers like the thought of dying.

> a closed door: all the information is there, only I don't have the key.

> the Sahara Desert: I wander about aimlessly, trying to find the right direction, yet always being fooled by mirages.

> quicksand: I find myself drowning in a mass of equations and variables, finding that the more I struggle, the more I drown.

These are some of the metaphors students chose for mathematics early in my Writing Seminar in Mathematics course. Their metaphors make it clear that the students who take the course enter reluctant to approach mathematics, seeing it as mechanical, rote, intimidating, and feeling powerless in its presence.

I offer the course at Ithaca College, a private, co-educational, four-year comprehensive college in upstate New York. The students are entering freshmen who are better-than-average writers, and who want to avoid a more traditional mathematics course. None of the students are mathematics or science majors. Most have interests in the social sciences or the humanities, but have not yet declared a major. The students vary in their mathematical background and confidence, but few believe that they have any mathematical skill or competence.

The goal ... is to change these students' conceptions of mathematics as a discipline and of themselves as learners of mathematics.

The goal of this paper is not to describe the Seminar so that the reader can reproduce it, but rather to entice you to try similar writing assignments in your own courses. I will provide five mathematical situations, each of which could be used in many different mathematical settings, and discuss how these examples, used in the context of cognitive-developmental theory, can become catalysts for changing students' conceptions of mathematical knowledge. Examples of students' writing, which illustrate their mathematical empowerment, will be shared as well.

The goal of the Writing Seminar in Mathematics is to change these students' conceptions of mathematics as a discipline and of themselves as learners of mathematics. I accomplish this by having them participate in mathematical situations, discuss their ideas both in small groups and with the whole class, write informally about their ideas and the process they used to generate those ideas in a journal, and then to express their ideas more formally in well-developed essays.

Everybody Counts: A Report to the Nation on the Future of Mathematics Education, released in January 1989, presents a plan for radical change in mathematics education from kindergarten to graduate school. The report stresses that "students learn mathematics well only when they *construct* their own mathematical understanding" (p. 58). In the Writing Seminar I try to create situations in which students can start with their own ideas and use these ideas to construct their own understanding. Their informal writing in their journals allows me to participate in this process.

The report compares mathematics to writing, stating that:

> In each, the final product must express good ideas clearly and correctly, but the ideas must be present before the expression can take form. Good ideas poorly expressed can be revised to improve their form; empty ideas well expressed are not worth revising. (p. 44)

Many mathematics students believe that they are wrong unless their work is perfect. They do not see mathematics as a place for revision and refinement. They are, therefore, reluctant to take the first steps to mathematical thinking.

I believe that mathematical thinking is stimulated by instances of surprise, of contradiction, of believing, and of doubting. Therefore, I provide experiences that focus on a series of mathematical events particularly designed to raise questions and generate mathematical thinking. Students are encouraged to explore the situations, generally in small groups. They realize quickly that I am a source of ideas, hints, and encouragement; but that I will not give them answers. I help these students become aware of their own experiences with mathematics, accept that they can and do have ideas and intuitions about mathematics, and validate these ideas and intuitions through the use of mathematical theory.

Many mathematics students believe that they are wrong unless their work is perfect. They do not see mathematics as a place for revision and refinement.

Students write about their experience with these events in a journal which I read and comment on every other week. My comments support the students' ideas and their struggle to make sense of the material, raise questions of clarification, try to lead students to see their misconceptions and consider alternative points of view, and invite them to think more deeply about mathematics. While I suggest journal topics to students, I also encourage them to use the journal in their own ways. Each student writes two two-page entries per week. I try to develop dialogues with the students urging them to respond to my questions and comments regularly. (For an account of one student's experience in using a journal in this course, see [4].)

The Developmental Process of Change

Students seem to go through a developmental process in changing their view of mathematical knowledge. What I am seeing as I read the words my Writing Seminar students write indicates that the first step in a changed belief about mathematics comes when a student begins to own his or her own ideas. This seems to involve acknowledging that mathematics was and is made by people. This happens before students become comfortable with the theory, the algorithms, or the procedures that are so necessary in mathematics.

This process is supported by cognitive-developmental theory. The students enter my course believing that mathematics is something that they can have no ideas about; that every question has an answer; that every problem has a solution. Teachers know and teach these truths. Students learn mechanically, following the rules precisely. The excerpts from my students' metaphors for mathematics listed at the beginning of this essay reflect this view, which is termed "dualism" by William G. Perry, Jr. [8, 9] and "received knowledge" by Belenky, Clinchy, Goldberger, and Tarule [2].

I create an environment in this course that helps the students become aware of both diversity and uncertainty in mathematics and gain a more personal, subjective view of mathematics. At this point, their own experiences become worth considering. They still believe, however, that knowledge is the product rather than the process of the inquiry; but they can now learn from the experience of peers, for peers as well as authorities (teachers, textbooks) have ideas worth hearing; also they begin to mold ideas, but often without being willing or able to defend them. This new phase is called multiplicity, or sometimes personalism, by Perry, and subjectivism by Belenky et al.

These students next become aware that intuitions may deceive; that some truths are truer than others; that truths can be shared and expertise respected. They become aware that it is the process of constructing knowledge that is important. They move away from subjectivism, learning to respect established procedures and their own insights, listening carefully to other points of view while evaluating their own. The course helps students move through this process in their view of mathematics. Through this process students become empowered and then become less fearful of the procedures and techniques needed to do mathematics.

Reclaiming Intuition

The following three mathematical situations are used early in the course to help students become more subjective about mathematics, and to acknowledge that they can and do have ideas about mathematical questions and that many of these ideas are valid. Students seem to reject their own thinking when they believe that mathematics must by learned in a rote way. These situations are designed to help students to reclaim their intuition and common sense about mathematics and to enjoy the sense of empowerment and confidence that they feel when this happens.

SITUATION ONE
The first problem I ask the students to think about is:

The Handshake Question: If six of us wanted to meet by shaking each others' hands, how would you envision the number of handshakes?

Please, before you read on, stop and think about at least two ways that you would respond to this question.

I plan this as an early experience for several reasons: with a little time and encouragement everyone has an approach; a formula does not usually come to mind; it doesn't look like a traditional school problem and so seems more approachable; and the meaning of "handshake" is open to interpretation.

Students think about the question on paper individually. I encourage thinking in their own style, insist that we all remain quiet so everyone can get an idea on paper, and urge those with one method to find a second or third. The discussion focuses on the variety of methods people have chosen. (See [3, pp. 20–21], for a few of the possible methods.) We get as many of the methods on the board as possible and as the list grows someone will ask which one is "right." This leads to a discussion of the virtues of different methods, which will be interrupted by someone wanting to know which answer is correct.

By this time we will have at least the following answers: $6 \times 6, 6 \times 5$, $(6 \times 5)/2$, and $5 + 4 + 3 + 2 + 1$. We discuss what each is counting. For example, in 6×5 each person initiates five handshakes, one with each other person. A pair of people will shake hands twice, since each of the pair must initiate a handshake. A discussion of the importance of definitions ensues, since some want to count handshakes this way and others do not. I do not encourage consensus, but rather have each student write the following journal entry outside of class:

Journal Entry: Give two methods you used to find the number of handshakes for 6 people. Give two additional methods discussed in class. Which of these methods do you like best for finding the number of handshakes for 6 people? Why? If you had to find the number of handshakes for 17 people, which method(s) would you use? Why? Build a table with a column for the number of people and another for the number of handshakes needed for them to meet each other. Use 1, 2, 3, 4, 5, 6, 7 in the number of people column. Look at the handshakes column. What are the next three numbers? Why? If you needed to find the number of handshakes for 100 people, how would you do it? Is there anything else you want to say?

This handshake activity could be used in any class where you want to discuss mathematics as a search for patterns, where you want to stress the need for definitions, where you want to encourage multiple methods, or where you need to generate the formula for the sum of the first n positive integers.

Students find it empowering to see so many different methods; for many the empowerment is to see their own method among them. Many students have told me that their mathematics classroom survival strategy has always been to wait for the answer to be presented. These students need to be encouraged to try to solve this problem for themselves; they will try enthusiastically when given "permission."

SITUATION TWO
To help students think more deeply about what they are doing when solving a mathematics problem or using an algorithm, I often ask them in writing to explain a concept to a close friend not in their class, to a parent, to a roommate, to a younger sibling, or to a sixth grader. I want them to use their own words, and to avoid mathematical jargon that is not their own. Try having your students explain a concept to a peer. You will learn what they really understand about the concept. This is an appropriate activity for any mathematics class. In the Writing Seminar, the assignment is:

Fraction Paper: Write a paper explaining to a sixth grader both how and why we add (or multiply) numeric fractions the way we do.

Students remember quickly how to add (or multiply) fractions, but are convinced that they were never told why and cannot figure it out. By working cooperatively in small groups, asking questions, and writing drafts in their journals, they develop good explanations.

They are again empowered through developing their own mathematical ideas and explanations, by understanding why we multiply fractions as we do, and by realizing that they have not all chosen the same explanations to use.

SITUATION THREE

I emphasize examples of the Fibonacci sequence and the golden ratio. Students are surprised that these numbers appear in situations that seem to be totally unrelated. They are also surprised, after counting spirals on pine cones, pineapples, and artichokes, that these numbers occur in nature. One of several problems that I use is Fibonacci's rabbit problem:

> *Rabbit Problem*: A pair of rabbits one month old are too young to produce more rabbits, but suppose that in their second month and every month thereafter they produce a new pair. If each new pair of rabbits does the same, and none of the rabbits die, how many rabbits will there be at the beginning of each month? Work the problem until you can predict the number of rabbits at the beginning of the twelfth month.

Students' Changing Conceptions of Mathematics

As I have indicated, my goal in this course is to change my students' conceptions of mathematics as a discipline and of themselves as learners of mathematics. Part of this change is to reclaim their intuition and common sense about mathematical ideas, an intuition which is lost when they come to believe that their own thinking is inappropriate in a mathematics classroom.

After a number of different mathematical experiences including the handshake question, the fractions paper, discovering the Fibonacci sequence and golden ratio in several settings, and reading *Flatland* [1] have been completed, I give the following writing assignment.

> *Change Paper*: Write a paper describing a specific moment when you realized a change in your thinking, in your mode of acting, or in your feeling about a mathematical question during a classroom mathematical episode.

Students are urged to develop their ideas informally in their journals before preparing a more polished paper.

One student, whom I will call Janet, chose her experience working on the rabbit problem as her moment of change. She first talks (in her journal) about the way she approached mathematical problems before the episode with the rabbit problem:

> Immediately before the rabbit episode, I tensed at the thought of doing a word problem. It's a mental block, almost. I see a word problem, and I automatically prepare myself for confusion. Usually when doing problems in class, I can follow the solution—as long as someone else does it first. (This applies to all problems, not just word problems.) So when approaching the rabbit problem, I figured that I'd attempt to do the problem, not be able to, and then wait for the answer to be explained to me so I'd have some idea of what was going on. I realize now this was a negative attitude to take, but then it didn't seem like an "attitude" at all—it was simply the way I had always approached mathematics.

Next Janet describes (in her journal) her experience working on the problem:

> What is the problem asking? What necessary information is given? At first the problem didn't seem too hard—each month a new pair of rabbits is produced from the existing pairs. The

number doubles every month. But as we started discussing the problem out loud in class, I realized that I was not allowing for the one-month-old non-reproductive stage. I tried working with pictures again, as was being demonstrated on the board, but it seemed to be getting too confusing for me to follow.

> I recalled that a chart had helped in the tiles problem, so I decided to make a chart. Making the chart was easy to do—the number of newborns the month before was the number of one-month-olds for this month; and the number of adults and newborns was always the same. As I worked on the chart, it was like a light shining through the leaves of a dense forest. (I know it's a lousy simile, but that's really how it was.) I saw. I knew. I wanted to raise my hand and shout out, "I have the answer! It's Fibonacci!" It seems really silly now, but at the time I was ecstatic.

Finally, Janet describes (in her change paper) her thoughts and feelings after the episode:

> I felt proud that I solved the problem alone, but something else increased my confidence. My friend, Ginger, who I consider quite intelligent, showed amazement when I solved the problem so accurately. She still didn't understand how I arrived at the solution, and she asked me to explain. Shocked, I told her that I probably needed more math help than she did. However, she insisted, so I attempted an explanation. As I explained the solving process, I found that I truly understood the problem. Helping Ginger understand it made me feel confident, not just about the rabbit problem, but about the whole mathematics subject. I never helped other people with math before; they always helped me. This new role gave me confidence that never before existed.

> After this event, math lost the mysteriousness and impossibility I thought it previously possessed. I lost some intimidation and frustration that I felt when dealing with math.

Janet expected not to be able to do the problem, but she did try a procedure, using a chart, that had worked for her small group on an earlier problem. Not only did she come up with a solution, but it was more satisfying and more efficient than the one being developed on the blackboard. Her sense of feeling empowered by her success overwhelmed her—first when she recognized she had a solution, and second when she realized she could explain it to someone else.

Students seem to reject their own thinking when they believe that mathematics must by learned in a rote way.

A second student, whom I will call Lee, wrote his change paper about his experience writing to a sixth grader about how and why we multiply fractions the way we do. First Lee describes (in his journal) his view of doing mathematics before he worked on his fraction paper assignment:

> Math is done in a strict, step-by-step process in which one missing, illogical, or wrongly done step will ruin the entire problem. Finding the correct answer in math is the main thing, while using the taught method to find that answer is also important. Doing mathematics can result in a precise answer or an estimate but it is not really a thinking process. Rather it is a process of identifying, comparing, and doing a problem in relationship to that identification and comparison. If you got the problem wrong there are no "ifs," "ands," or "buts"; you got it wrong. If you got the problem right and used the correct

method, you got the problem right. The math process is one in which all attention is focused on a narrow subject. My mind is not allowed to create, wander, or think about doing the problem. My mind says, "Compare this problem to others that are like it and base your answer on the way you found your answer to that other problem."

Next Lee describes (in his change paper) his experience working on the assignment:

We worked in small groups first and I extended this group activity to my individual work. Surprisingly, the dreaded "identifying and comparing process" did not appear in my figuring of the problem. I had no example to follow. I could not merely solve the problem using a previously specified method.

I like writing mainly because I receive satisfaction when I finish a solidly written piece. While writing I am, at the same time, thinking and understanding what I'm writing about. Until the fractions paper I could not say the same about doing a math problem. I thought out and understood what I was doing while I was figuring how and why we multiply fractions the way we do. I felt like I was writing my own chapter of a math textbook.

Finally Lee describes his feelings and thoughts after the episode with the fractions paper. The first quote is from his change paper and the second is from his journal.

My doing the fractions problem well produced a satisfaction that sparked my interest in math. Furthermore, because I did the problem without referring to any examples of any supposedly proper methodology, I controlled the problem. I chose to take the problem in the direction I did. Finally, the ability to be creative made me want to do the problem and do it well.

Math can be more than just identification and comparison. Math can involve letting the mind wander, create, and yes— think. My mind can say, "Well why don't I try this another way so I can draw up new conclusions or relationships." Math can include more than just numbers and formulas.

I like writing mainly because I receive satisfaction when I finish a solidly written piece. While writing I am, at the same time, thinking and understanding what I'm writing about. Until the fractions paper I could not say the same about doing a math problem.

For Lee mathematics had been a subject in which he did not think, but rather identified and compared. As he worked on the fractions paper, he realized that he would not be given a model; that, indeed, he was being asked to think the problem out for himself. To think, to develop his own ideas, to come to understand the process of multiplying fractions in his own way were empowering and satisfying experiences for Lee.

Beyond Intuition

The early situations in the class are designed to get students to accept and reclaim their intuition and common sense about mathematics, an important step in changing their views of mathematics. However, in mathematics one's intuition must be supported by theory, logic, and data. The theory, logic, data, and the intuition must be evaluated.

After students have gained some confidence and courage to try mathematical questions and to share their own mathematical ideas, I use the following question, The Belted Earth, to generate a conflict between their intuition and the mathematical theory developed to solve the problem. This question helps make students aware of this conflict and urges them to resolve it for themselves.

SITUATION FOUR
Please, before you read further, read the Belted Earth question and resolve it to your own satisfaction.

The Belted Earth: In the following question focus first on your initial intuitive response, only then think about a solution.

Think of the equator. Put a flexible steel belt around the earth at the equator so that it follows exactly the contours of the earth. Now add forty feet to the length of that belt and arrange it so that the belt is above the equator for its entire length. The belt still follows the contours of the earth at the equator and is raised above the equator by the same distance at every point.

The question is, what will fit between the earth and the belt? That is, what is the distance between the earth and the belt?

Most students (in fact, most people) believe only something very tiny will fit under the belt and are surprised by an answer of more than six feet. Following an extensive discussion of both intuitive and algebraic solutions, students are asked to write the following journal entry.

Journal Entry: Describe your intuition about the Belted Earth question. Discuss the theory developed in class. Which one do you believe? How are you resolving the disparity between your intuition and the solution developed in class?

I read and comment on this entry extensively. Students now have an investment in understanding mathematical theory and want to make sense of this surprising result. They want to understand what pi is, why we don't need the actual circumference of the earth, and why adding 40 feet to any circumference yields the same result of more than six feet. See [3, pp. 22–23] for a discussion of the development of this problem and the responses to it by a group of intellectually able, adult women.

SITUATION FIVE
A second situation designed to help students acknowledge their intuition and then look to theory, logic, and data to evaluate it, is the extensive mathematics unit developing the notion of "straight on a sphere." Students begin with their intuition about "straight" on a curved surface, but then are encouraged through the "believing game" [6, pp. 147–191] to consider the perspective of someone else who might have a different belief than they do. They participate in the development of some mathematical theory and at least one mathematical proof, and develop in their journal writing their understanding of these mathematical ideas. The following describes the sequence of activities that form this unit.

STRAIGHT ON A BALL

First, in one class we discuss the many uses of the word straight in our general language and in our mathematical language. I then ask the students to consider whether or not we could have a straight line on the surface of the large rubber ball I hold before them. We have a lively discussion with many ideas both pro and con. Then I ask them to write a journal entry explaining what straight means to them and discussing whether or not we could have straight lines on the surface of the ball (or the earth).

Second, I read and comment on their journal entries about straight on the earth. My comments primarily ask for clarification, especially when some of their statements are contradictory. Students tend to discuss one or more of the following:

> Straight applies to lines on the plane and not to lines on curved surfaces.

> Straight is the shortest distance between two points and so can exist on a sphere.

> Straight is a matter of perspective. A student will talk about a straight line on a sheet of paper, and ask if it loses its straightness when the paper is bent. Another student will argue that if you look directly at the equator drawn on a globe, it will look straight, but if you turn the globe slightly, it will appear curved.

Third, I ask them to reread their own journal entry and then in small groups to play the believing game [6, pp. 147–191], believing that there are straight lines on the surface of the ball. While many do not hold this belief (a straight line is defined on the plane!), they work hard considering what a person who does believe it would have to say. (For example, the person might say that there must be a shortest distance between points on the sphere.) Following the small group and whole class discussion of these beliefs, we agree to define great circles as "straight" lines on the surface of a sphere.

Fourth, we discuss the development of spherical geometry as mathematicians have described it, going back to Euclid's postulates. We see triangles with an angle sum greater than $180°$. You can watch the ripple of surprise/disbelief move through the class when you produce (using rubber bands on a ball) a triangle with three right angles clearly defined. We then consider properties we retain and properties we lose as we move from geometry on the plane to geometry on the sphere. We develop the formula for the area of a spherical triangle using Jeffrey Weeks' method [10, pp. 137–148].

Fifth, in their journals, I encourage the students to consider what might be "straight" on other surfaces, like a cone, a cylinder, or the walls of our classroom.

Sixth, in their journals, I ask them to consider why they learn only Euclidean geometry since we do, in fact, live on a surface closer to a sphere than a plane. I ask them to make that consideration in light of what we have discussed as a class and of their own experience.

Seventh, I ask them to consider the historical development of spherical geometry in response to the question, "Was spherical geometry invented or discovered?"

Lee was reluctant to accept the idea that "straight" could apply to lines drawn on curved surfaces. We developed a dialogue in his journal concerning "straight on a ball" that began when I had trouble understanding where he stood on the question and needed clarification.

DOROTHY.   Can you have straight lines on the surface of the ball?

LEE.   No. It is impossible to have straight lines on the surface of a ball because even the slightest of curves cannot accommodate a straight line. There is no part of the surface of a ball that is not curved. There is the possibility that a small marking on a ball may appear like a straight line. But this is only appearance. Sometimes the human eye alone cannot perceive accurately between straightness and roundness.

DOROTHY.   Invent a new word for the thing you would consider the closest that you can come to "straight" on the surface of a ball? What properties would these things need to have?

LEE.   The closest thing to "straight" on the surface of a ball is called "stirct." If something is stirct it must follow the shape of the round surface perfectly. The stirct line will meet itself across the surface of a ball an infinite number of times. The stirct lines can have no deviations from this infinite cycle. A stirct line is the same as a straight line in all things except that it does not have to exist on a plane and that all of its points will meet.

Lee was content to talk about great circles as we developed the theory of spherical geometry in the class discussion. He was, however, reluctant to use "straight" in that discussion. In his journal I continued to ask him to develop and refine his definition of "stirct" by considering lines on a cone, a cylinder, and the walls of our classroom. At the end of the semester, he was still developing these ideas, but had made his definition of "stirct" less restrictive.

Metaphors for Mathematics

In this paper I have described three mathematical situations (the handshake question, the fraction paper, and the Fibonacci sequence and rabbit problem) that help students to reclaim their intuition in mathematics and two mathematical situations (the belted earth and straight on a ball) that help students to move beyond their intuition and use theory, logic, and data to support their ideas. I have discussed how I participate in their developing thinking as I read and respond to their writing in their journals.

I see the signs of change in their view of mathematics as I read their journals, but I am not always prepared for the descriptions of empowerment that are documented in their change papers. The excerpts presented here in the words Janet and Lee used to develop their change papers are typical descriptions of the empowerment students feel through their changing conception of mathematics.

For me math is most like the Sahara Desert. I wander about aimlessly, trying to find the right direction, yet always being fooled by mirages.

I have one other indicator of students' changing mathematical views during this course. Early in the semester I gather the students' metaphors for mathematics. I usually wait until the second week, because a trust relationship needs to begin to develop before the students will be candid about their views of mathematics.

> *Metaphor.*   In the classroom setting each student writes his or her own responses to each of the following orally given directives:

> List all the words you would use to describe mathematics. (5–7 minutes)

> Imagine yourself in a situation of doing mathematics and list all of the feelings that come to mind. (5–7 minutes)

> List all of the objects (nouns, things) that math is like for you. (5–7 minutes)

> Read over your three lists and write a paragraph beginning, "For me math is most like a . . ." (20 minutes)

On the first page of this paper are excerpts from the paragraphs written by Writing Seminar students. I collect this first set of metaphors; then, about the twelfth week (of a fourteen week semester), I gather metaphors a second time. The following class I return both metaphors to the individual students and ask them to write the following journal entry reflecting on their metaphors and the similarities and differences they observe about themselves as they read the metaphors they have chosen for mathematics.

Journal Entry. Place the two metaphors that you have written this semester in your journal. How are they similar? How are they different? How does each one of them fit your present view of mathematics and of yourself as a learner of mathematics?

Is there a way you would like to view mathematics that is not yours yet? What metaphor would you choose to describe this new view for yourself? What experiences would you need to be able to adopt this new view?

Janet's Changing Metaphors

To conclude, let me share the metaphors that Janet used to describe mathematics as the semester progressed. Her first metaphor has characteristics that I often see. As she wanders aimlessly she is at the mercy of the elements, and a sudden, blinding sandstorm can destroy any progress she has made:

For me math is most like the Sahara Desert. I wander about aimlessly, trying to find the right direction, yet always being fooled by mirages. Occasionally I might stumble upon a lake filled with ice cold water, but soon enough I get lost again, surrounded by the expanse of desert. The intense heat gives me a headache as I try to remember where I came from and where I'm going. All the landmarks look the same—each cactus, sand dune, and lizard. The wind picks up and I can find myself in a sandstorm, the sand blinding me totally, making me oblivious to everything else. Once it clears all the knowledge I had gained now is lost, for I'm in a different part of the desert, once again unable to find my way to solid land.

In her second metaphor Janet indicates that she often still feels out of control and still experiences illusions, but she is now on a set course and she can experience exhilaration, fun, and challenge:

For me, math is most like a roller coaster. Math has its ups and downs. Sometimes there are these really big hills that seem to go straight up forever. It's really a slow, arduous climb to the top. Once you get to the top, there can be that exhilarating feeling of flying down the hill at top speed without a care in the world. At other times, getting to the top is only an illusion. The "top" turns out to be level ground with another huge hill ahead. A roller coaster is also a huge loop, a circle with no beginning and no end. The ride can be fun and challenging at times; yet at other times it can be dark and scary.

Janet, in her journal, talked about these two metaphors:

I guess I feel that I'm no longer totally <u>helpless</u> when it comes to math. I've come to realize that math is not <u>all</u> bad, as the desert metaphor suggests. There are challenging aspects to math, and if you are courageous enough to brave the big hills, you may be rewarded by a good ride down.

Another difference between these two metaphors is that in the Sahara Desert I was all alone, but on a roller coaster there are other people to go through the experience with you. I've discovered that I'm far from alone in my frustration with math. And yet, I've seen glimpses of how some people are fulfilled by the challenges that math offers.

In a desert, you can walk for hours and not know when you'll get where you are going. At least on a roller coaster, there's a set track and it's the ride that's difficult.

Janet describes her ideal metaphor for mathematics as the following:

—a forest with patches of sunlight but shade from the trees. Everything is different—different kinds of trees, different kinds of leaves, different kinds of animals. You might get lost sometimes, but you're never scared. Even in the darkest part of the forest the leaves will never be so dense that you can't see the sun.

Remember her "lousy simile," "it was like a light shining through the leaves of a dense forest?" (See page 3 of this paper.) This metaphor is not new to her thinking!

Janet realizes that to adopt her ideal forest metaphor she would need to change her attitude:

Taking a positive attitude towards math—not saying this is going to be hard for me; this is always going to be hard for me. But saying this is a challenge and I'm going to work at this until I get it.

Janet, who began the course talking about doing mathematics as helplessly wandering in a desert, describes how she now sees doing mathematics in the context of her forest metaphor.

In doing a problem you think of a solution that is correct, but you aren't satisfied with your first solution; you just keep working on it. Being comfortable enough with a concept to want to keep working with it, to find different paths to a different solution. Just getting the right answer is not the ultimate goal if you want to be in the forest. You want to explore things, not just solve things!

Janet was empowered when she developed a procedure to solve the rabbit problem. She now wants, through mathematics, to explore things, to be unsatisfied with her first correct answer, but to enjoy the challenge of looking for alternative paths, methods, and possibly even alternative results.

In the Writing Seminar I use the metaphors to monitor changing attitudes and to help the students acknowledge their own changing conceptions of mathematical knowledge and mathematics learning. In other settings I gather metaphors in one class, read them, and return them the next class, generating a discussion about mathematics and mathematics learning. This discussion can change the classroom environment to one in which students are more open in their thinking and more willing to try their own ideas. See [5] for the metaphors of some secondary school Advanced Placement Calculus students.

Writing opportunities ... have put me in touch with students' conceptions of mathematics ... The candid writing of students continues to surprise, delight, distress, sustain, and teach me ...

Writing opportunities, like those described here, have put me in touch with students' conceptions of mathematics and made me aware of their survival strategies for the mathematics classroom. Through nontraditional mathematical situations, involving writing, I have been able to change students' disempowered patterns and approaches to mathematics learning and help them feel the empowerment that comes through reclaiming their mathematical intuition and testing that intuition against mathematical theory. Cognitive-developmental theory has helped me understand this change process and guided my design of writing opportunities to facilitate this change. The candid writing of students continues to surprise, delight, distress, sustain, and teach me, as I work to empower students in mathematics. I invite the reader of this essay to try these writing opportunities and others of his or her own design. Begin slowly. Read carefully and seriously what your students say. Ask for clarification to help you really understand their thinking. Be prepared to be surprised. Be patient both with your students and with yourself. You will find that both you and your students will become empowered by the experience.

REFERENCES

1. Edwin Abbott, *Flatland*, Dover, New York, 1952.

2. Mary Belenky, Blythe Clinchy, Nancy Goldberger, and Jill Tarule, *Women's Ways of Knowing: the Development of Self, Voice, and Mind*, Basic Books, New York, 1986.

3. Dorothy Buerk, "An Experience with Some Able Women Who Avoid Mathematics," *For the Learning of Mathematics*, 3(1981): 19–24.

4. ——, "Carolyn Werbel's Journal: Voicing the Struggle to Make Meaning in Mathematics," Working Paper #160, 1986. Available from the Wellesley College Center for Research on Women, Wellesley, Massachusetts 02181 for $3. (ERIC Document Reproduction Service No. ED 297 977, microfiche only.)

5. ——, "Mathematical Metaphors from Advanced Placement Students," *Humanistic Mathematics Network Newsletter*, #3, 1988. Available from Alvin White, editor, Department of Mathematics, Harvey Mudd College, Claremont, California 91711.

6. Peter Elbow, *Writing without Teachers*, Oxford University Press, London, 1973.

7. National Research Council, *Everybody Counts: A Report to the Nation on the Future of Mathematics Education*, National Academy Press, Washington, DC, 1989.

8. William G. Perry, Jr., *Forms of Intellectual and Ethical Development in the College Years: A Scheme*, Holt, Rinehart, and Winston, New York, 1970.

9. ——, "Cognitive and Ethical Growth: The Making of Meaning," in A. Chickering (ed.), *The Modern American College*, Jossey-Bass, San Francisco, 1981: 76–116.

10. Jeffrey Weeks, *The Shape of Space: How to Visualize Surfaces and Three-Dimensional Manifolds*, Marcel Dekker, New York, 1985.

Two Perspectives on a Writing Intensive Course in Operations Research

Part I: Not Just Another Class

Mary Margaret Hart McDonald, student
Radford University, Radford, Virginia

I don't clearly remember the purpose of choosing Stat 441, Introduction to Operations Research, as an elective in my schedule. In fact, the previous year, I really had no idea statistics existed as a major area of study. But after taking a statistics course, I found that I really enjoyed statistics. So I decided to get a major in it, along with my major in mathematics. I had heard of operations research as a branch of statistics, and decided to give it a try. After all, I came to Radford University to further my education.

So I entered this class with an open mind, a good thing, because the course was designed to use experimental learning techniques. The students were going to write in this class for a grade!

I had never imagined being given writing assignments in a statistics course (or related field). I suppose I decided to major in these areas partially to avoid such assignments. I was quite surprised when we received the syllabus on that first day of my senior year of college. Our grade would be based on not only homework assignments and three in-class exams, which would come from our book and class notes, but also informal journal writings and a formal project to be written and presented orally. A "typical" statistics course includes being taught a technique and mimicking its use (not writing about it). Too often students do not think of the applications. Learning how to do "it" and remembering "it" for the test are the most important things to a "typical" student in a "typical" statistics course.

I found there were no right or wrong ways of finding ... answers. We would think and therefore learn through our writing.

I did not consider this course, even on the first day, to be a typical course, and neither did I consider the four students and teacher involved with the course as typical. But for some reason, I felt differently about this class. Maybe it was the first time I was scared when looking at a syllabus. Maybe it was because this was not a required course. But whatever the case may be, if it had not been a 400 level course in my major, I do not think that I would have stuck with it and become so dedicated to the learning process. I'm glad I did, because something valuable was to be learned by the time I had completed these 3 of 126 credits that I would accumulate over a four-year period. I wanted to learn from this class, although I was not exactly sure what it was that I expected to learn. But by looking at the syllabus, I knew I could not take this course lightly, as was the consensus of the entire class. Of course, if we had known the requirements for the course before registration, we all might have registered for a different statistics course that semester. In retrospect, I am thankful for having to do all that work, and I would recommend the class to those dedicated few who may ask me about taking the class. I would tell them that taking the time to complete the assignments is well worth the sacrifice.

Free Writing

The new thinking technique which developed in this course, and in many classes involved with Radford's Writing Across the Curriculum Program, is called free writing. We were encouraged to reflect on the topic of the day and put the course into some kind of perspective in our journals—if that was possible. It helped to have a sample journal entry (see Appendix) in the beginning since this approach was so new to these four senior undergraduate students. We spent some time reading over this entry and discussing how this was not the only way to write in our journals, but that we should try to think about the new material, connections, and applications of it in our writing. I'm sure every individual developed his/her own method of doing this new type of work. There is a sense of security in knowing what is expected from you, and yet knowing that you have freedom to think, not analytically, as we are used to in mathematics, but in an individualistic style. There were no right or wrong answers, and I found there were no right or wrong ways of finding those answers. We would think and therefore learn through our writing.

Each assignment was, in some ways, intriguing. Our first in-class journal writing was to be about "What the World Needs and Why." Now it isn't often that this is how a course in such a predictable discipline will start. This assignment, as well as others, served a purpose, actually, many purposes. The purpose of this first entry did not become clear until later in the semester. It was to make us think about a topic for our major project. In a nutshell, this project we were required to do was to take a real life situation that seems to pose a problem, and try to solve it using the techniques we would learn in the class.

In fact, a variety of topics were assigned for us to write about including "Situations on Campus Life" and "Things that would be Useful to be Optimal" and then "On a Particular Situation" to help us decide which problem we would tackle that would be worthy of a semester's work. After our in-class writing on optimality, we had a class discussion in order to compile a list of possible project topics. Our assignment that night was to pick one thing from our list and write about it. One student had thought of scheduling at the airport. Since I had, at one time thought of pursuing Air Traffic Control as a profession, I chose to write on this topic. I discovered just how little we knew and how broad the topic was. I found myself commenting that

> A lot of research would need to be done ... It would be interesting I think, so I would want to work on the project the whole semester—I definitely want something that would hold my interest.   (September 9, 1987)

Through writing, we continued to investigate different options until we pinpointed just one situation that we wished to handle. After a week and a half, both the students and the teacher agreed upon a topic of snow removal in the City of Radford. I think creative writing played an important role in having the class recognize possible project topics. It really got us thinking. The style used on that first day was indicative of the rest of the semester: we were to be thinking for this class, not casually copying notes, problems, and passages from other material.

Communication

The small class size was a special feature that, I feel, made the class work especially well. With only four students, it was possible to be required to do more work and both the students and the teacher could get feedback from each assignment. The teacher knew where the class stood, when in journal entries she saw comments such as

Okay, I'm still confused!   (December 4, 1987)

and

Again, I'm quite puzzled.   (December 7, 1987)

Or even on the few occasions when an entry read

Oh Boy! We're starting some good stuff now! Networks, Cycles, Trees, Paths, etc. this all makes sense! This is great! You wonder why at this point, why people don't like Math/Stat— why they say it doesn't make sense! This stuff is so logical!
(October 7, 1987)

and

Project scheduling sounds pretty interesting. Again, it's a real-life problem that we're solving. But it does take a while just like the rest of the stuff we've done. It all makes sense too.
(November 11, 1987)

The students felt comfortable writing to the teacher. It was much like a conversation, but better. Many students are too shy to say anything in class, or even one-on-one with the teacher about general problems, concerns, or even praises. Students feel a piece of paper and a pen can protect them by removing them ever-so-slightly from the teacher. A false sense of security maybe, but a useful one indeed, one that our class began to count on. Feedback is also useful to students because they know that the teacher cares when they see comments on their papers like

Good idea!   (September 18, 1987)

NICE thought and perspective.   (November 13, 1987)

This really became a horribly long lp [linear programming] problem—NOT one I would assign you!
(November 16, 1987)

Feedback is important because when the student knows the teacher cares, the student is more likely to put some extra effort into the assignments.

Development

Through those first eight weeks of that semester, my goal was simply to get the journal assignment done. I do not feel that such an assignment is done adequately by sitting down the night before the journal is due, trying to record what happened in several previous class periods. It is achieved by taking the journal assignment seriously, even if for no other purpose than to put serious thought into it for the grade. My writing at the beginning of the semester was not filled with thoughtfulness, just declarations like

It seems like it [the Simplex Algorithm] is a reasoning process— I don't know much about it to this point. I suppose I'll be able to use it and more of why we'll use it after Monday.
(September 16, 1987)

I want to see an example to see what I'm really supposed to see through the homework.   (October 2, 1987)

which showed just how mechanical I had learned to become over the years. The motivation for writing should not be the focus here though, because no matter what the initial motives, serious thought will produce something worthwhile in time. What became worthwhile to me was learning how to think. I might not have become aware of this were it not for the requirement to write about what we were really doing in that class. And it took a while for me to see the potential, but it was fascinating when I did. I began to piece together what we were doing without being told how that was done.

The transportation problem deals, basically, with getting something moved from one place to another, with several sources and several destinations. In our project we have starting places (sources) . . . We want to get from the starting point to various intersections (destinations) . . . We have ____ as a supply, ____ would be our demand . . . The cost involved would be assigned as total cost . . . we could compose a table . . .
(October 26, 1987)

This [Project Scheduling] can be used for any kind of project that we do.   (November 11, 1987)

Another benefit achieved through informal writing is the improvement of good communication skills, a necessity in the world that we must enter upon completing the required number of classes at college. This course was designed to familiarize us with practical statistical applications that we might encounter in the "real world." Improved communication skills in our simulated real world was exactly what we developed, to our surprise. Just a few weeks before the end of the semester, the class applied these writing techniques to our project, without having been assigned to do so. After sitting around Jim's kitchen table for 45 minutes wondering how we would start our formal report of our project, Jim suggested we each take out a piece of paper and start writing. So we did. After about five minutes he called time and we each read what we had written. It was amazing to note that we had found our starting place within ten minutes of having read our work.

The applications of free writing are endless. In addition to getting us started, our journals also led us to discover new uses for a technique. An example of this was found in my journal writings which began with: "We did the assignment problem today" and halfway through the entry say: "I wonder if it could be twisted to the snow plow problem."

Informal journal writings allow the student to discover. They can find answers, not be given answers.

Students can also answer questions many times on their own, rather than depending on the teacher for answers. This ability to discover can be not only beneficial to the student, but also to the teacher. We too often find ourselves depending on the teacher to answer our questions. The students find themselves powerless in a classroom situation against a teacher who is perceived as being "all-powerful" and "all-knowing." Informal journal writings allow the student to discover. They can find answers, not be given answers, as was evidenced in my writing.

Why did we decide to do the homework problem by dynamic programming? We can do the Capital Budgeting Problem as linear programming, or by assignment, as said in class. Now we did it by dynamic programming, and there's probably several other ways to do it. How do you figure out which way is the best? . . . by doing it like this you don't have to draw huge charts, or put equations into the computer. You don't always have a computer available. And when doing it by hand, if it looks too complicated, you can always resort to a different method. It helps to know that you don't have to stay stuck. There is a way "out."   (November 30, 1987)

So with this freedom of writing, the teacher need not answer all questions, as the students find their own answers. A sense of confidence also comes with this discovery which seems to make up for the feelings of frustration that the student holds.

Reservations

The end of the semester came quickly to the surprise of the class. Although it had not been stated exactly what topics would be covered that semester, I felt it ended without having covered as much material as I had expected. Class time seemed to fly by on many days without my having learned anything new about operations research. In fact, on many occasions my journal entries began with

Well, we didn't really get anything else done today ... this class just goes by so fast! It seems like we don't ever get anywhere ...   (September 25, 1987)

This is a blessing to many students, but not to me, at least not in such a practical class. I now worry about the strength of my background in operations research compared to the background of other students who have had a traditional (non-writing-intensive) course. My fears lie in not having covered as much material as these other students, because we spent time writing rather than being lectured to about operations research techniques. Even though I feel my exposure to the subject material is limited, I believe that when I reference this material in the future, I will recall it faster due to the thinking process developed. And in the case that I have not seen the material before, I will be able to learn it faster than my peers. But of course, this cannot be verified until the situation arises, which may never be.

Recommendations

The value of free writing was picked up by the class mostly because we were required to do the writing regularly. By taking up collections of choice entries at irregular time intervals, the class kept up and eventually learned the value of free writing. Because not every entry can be collected and taken seriously in a big class, entries must be collected only at certain times. I feel that a "surprise" collection rather than having preset dates for collections keep the students "on their toes." It is, for many students, just getting the bare minimum done that is important. So when students know that every entry isn't going to be collected every day, they won't do them every day. And if they know what entries are to be collected and when, these are the only ones which will get done. I think a varied schedule for collection would encourage the students to keep up with their journals, and in the process, they may reap the many hidden benefits.

It also helps to occasionally be given topics to write on because we all do get stuck sometimes. Some topics given to us were on "Techniques," "How to Start a Problem," "Why we were doing the work by hand when it could be done on the computer," "Possible Test Questions," "Advantages and Disadvantages," and many more. Even more than giving us something to think about, we were given some direction, something students are used to having and many students don't know what to do without.

Discussions of our writings kept us unified. We were forced to discuss what we had written, but the manner in which we did that was up to us. We could read, summarize, or use a combination of both. We needed to be forced to talk about what we had written. And we always seemed to find out, after the discussion, that what we had written was not so silly after all. I think that this is the biggest fear in why students will not talk in class. They are afraid of being laughed at by their peers. By writing down our thoughts first, we were forced to acknowledge that we had thoughts. Through discussion of our entries, we realized that we did know something, and that's important to a student. The thoughts were now not just thoughts written down that we would never look at again, but they were concrete ideas that could be referenced in the future and be used in solving our homework or project problems.

Epilogue

The student/teacher dedication/enthusiasm exchange is improved through the cooperation of students with teachers. Teachers must be willing to work with the students, for the students. In return, students need to take their education seriously, and complete assignments with thought. For example, if teachers would keep from assigning "busy work," then the students would be able to find purpose in assignments. When this purpose is evident, students have more motivation to put time and thought into them. They can then get excited about going to class and learning. Seeing this excitement in students' eyes, I would hope, would excite the teacher. And the exchange is complete. Properly utilized teaching techniques, including writing in nontraditional writing courses, can spark enthusiasm. It did for me. I am thankful that I took advantage of the opportunity I was given to develop my awareness of my ability to learn; and to further encourage my professor to continue to teach others the same.

Part II: A Writing Project in Operations Research

Coreen Mett, professor
Radford University, Radford, Virginia

For more than three years I have used a variety of writing to learn devices in all levels of mathematics classes. Journals have proved their value as a place for students to organize, to explore, and to reflect [1]. In addition, an occasional 5 minutes spent writing in class, describing for a classmate some example, some application, or some procedure, have shown rewards in the increased responsibility students take for their own learning.

But until recently, none of that writing had been the typical "term paper" that is commonly thought of as writing in a course. A complete paper developed in a mathematics course seemed too distant from the immediate goals of a course. I hesitated to give any assignment which delegated to me the task of teaching writing—clearly the job of the English department.

This paper was associated with a semester-long project which would be centered on a problem found in campus or community life.

However, a course in operations research is the natural setting for a more "formal" type of writing. The study of operations research techniques includes many realistic problems where a correct answer does not necessarily exist. Therefore a good portion of time is spent designing an appropriate mathematical model, and justifying results. This course provides an opportunity for students to experience the reality of spending a large portion of work time in writing proposals, reporting progress, and validating assumptions. In order for a term paper to serve a purpose beyond some contrived course requirement, to reflect a typical working environment, this paper was associated with a semester-long project which would be centered on a problem found in campus or community life.

Course Description

As shown in the course syllabus in Appendix 1, about 30% of the final course grade was allocated to writing in nearly equal portions of informal writing, prospectus writing, and final report writing. The informal writing was a combination of in-class free writing and out-of-class journal writing. The journals were primarily a place for students to reflect, to summarize, to explore, and to communicate concerns. Mary Margaret Hart McDonald, a student in this class, describes in her paper, "Not Just Another Class," her reactions to the informal writing.

The monthly prospectus assignments were designed to be a series of progress reports, which would hopefully lead to a final project report. They were intended to parallel the kind of reporting done on a typical long-term project, informing sponsors of progress and outlining plans for further development. These reports were also meant to force students to grapple with their accomplishments and their goals in a manner which would keep them on schedule with the long-term problem.

And the final report, both written and oral, was a group effort. As with the prospectus, this design reflects the normal consulting atmosphere. Usually a team of people with various specialties goes through cycles of collaborating on general plans, then breaking away to pursue special aspects of the problem, and finally writing and presenting a single comprehensive set of results. I hoped to simulate this situation as closely as possible with our semester project. The fact that the project grade would be shared by all again represented a realistic responsibility. At the same time, the individual prospectus allowed me to detect unbalanced participation in the group and make credit adjustments, if necessary.

Informal Writing Forms the Link

One of my goals in the course was to ease anxiety over that major project and final paper by writing informally about the project. For example, at the very beginning of the course free writing was used to explore interesting possibilities for the semester project. First students were asked to write non-stop, without any concern for correctness, as quickly as possible about "What the world needs now." During another class they listed things around campus that needed improvement, and discussed problems they encountered in every day life. After only a week and a half of writing, searching, discussing, and thinking, the class was able to select the problem of improving the Radford City snow removal system.

After the students scheduled their first meeting with Mr. Farlow, the Radford City snow removal foreman, we prepared for the interview with a short informal writing. During class students were asked to "Imagine that you are the *only* person delegated by the class to meet with Mr. Farlow and learn about the Radford City snow removal procedure. What do you think he imagines you to be like?

What do you imagine him to be like? How will you begin the conversation in order to win his full cooperation? Formulate the two most important questions you feel you want to ask?" One student wrote:

> I would hope that he hasn't stereotyped me as the typical college student—the always partying type. This being a higher level class I would hope that he sees me as a mature, hard working and serious student. My image of him is a little blurry—kind of like a father figure. Is he bored or does he have more jobs than just snow removal—I should think so. I would want to start the conversation with an introduction and an assurance that I am willing to work with him in any way. I want to come across as serious and with the attitude that he is doing me a favor by seeing me and spending some time with me to talk about the project.
>
> The two most important questions—Do you have different plans for different weather conditions—in other words is there just one plan for every situation that arises? Will you be willing to use a new plan if we think it is more effective?
>
> (Hicks 9/21/87)

Subsequent discussion revealed common concerns over possible misconceptions about college students, and resulted in a plan for appropriate dress for the interview. Both the students and I felt confident as a result of this preparation for their first meeting with Mr. Farlow.

With the snow removal project in mind, I reorganized the standard topics on operations research (such as linear programming, network analysis, project scheduling, dynamic programming, etc.) into a more useful sequence. While we progressed through these techniques, we simultaneously developed our special topic on a parallel track. Since each new method might be utilized in examining the major problem, I tried to weave the possible connections with informal writing assignments.

For example, when we studied linear programming, students were asked to write and think about how they might pose a snow removal problem as a linear programming system. They were asked to "Define some variables, an objective function, and some constraints which might describe a snow removal problem." I wanted them to create, imagine, or design a maximization or minimization problem with some limitations related to a snow removal problem. I didn't care whether they wrote a paragraph of analysis or whether they scribbled a series of equations and formulas. But I hoped that they would designate some relevant variables and describe some constraints which would help them consider the fundamental aspects of the snow removal problem. Exactly what was to be decided? What were the assumptions? What were the factors influencing efficiency? What was the primary goal?

I was delighted with my own insight when I carried out this assignment along with the class. But I was disappointed in their results. They seemed bound by traditional problem-solving, unwilling to imagine or create their own new problem. So I shared my own writing with them, complete with a model:

minimize $z = \sum_{ij} x_{ij}$ where x_{ij} = amount of time plow i spends on road j subject to the constraints:

Estimated Time: $0 \leq x_{ij} \leq b_j$ where b_j = estimated time for road j

Priority: $x_{ik} \geq x_{il}$ where k represents a priority road and l represents a secondary road

Chemicals: $\sum_{i=1}^{4} a_{ij} x_{ij} \leq C_j$ where C_j = amount of chemicals allowed for road j and a_{ij} = rate of dumping chemical by plow i on road j (assumed to be a constant rate)

(Mett 10/2/87)

When we discussed my model, students noted that my priority constraint does not adequately represent the condition that primary roads be completed before secondary roads are begun. My constraint incorrectly requires that each secondary road be allocated less time than any primary road. But my mistake was beneficial in two ways. First, it emphasized that this writing is tentative, exploratory, not necessarily perfect, and that I felt comfortable with being wrong. Second, my error gave us the chance to discuss how we might correctly indicate priority with linear constraints. Most importantly, our discussion helped focus on a reasonable model of a snow removal problem by considering our objective to be minimizing the total time needed to clear all roads while considering limitations on chemical supply and concerns about priority. This began a debate on whether the goal of the project ought to be minimizing distance (by considering a shortest route) or minimizing time. A natural question arose over the equivalence of the two problems:

■ minimizing the time needed to cover a fixed distance;

■ maximizing the distance covered in a fixed time.

On the down side of this example, I must admit that my students appeared to think I was crazy to make up some artificial problem whose solution was therefore irrelevant to the problem they were really working on. I worried about their reactions. Would they assume I wanted them to solve the snow removal problem by forcing it to be a linear programming problem? Would they drop their own investigation to pursue my version? Would they feel I expected them to come up with a model like mine during a free writing session?

About two weeks after the linear programming experiment, I again tried using writing to form a connection between network problems and their major project. During class we all wrote freely on "Describe some similarities and differences between your snow removal problem and the standard 'shortest route' problem we have considered." This writing brought much more reflective results:

The snow removal problem is a basic shortest route problem. We want to find the shortest path for each driver to take. The differences start when you take into consideration running out of gas and running out of chemicals. We then need to find the shortest route to refuel and/or reload the chemicals.

(Hicks 10/14/87)

Another student showed an understanding that the setting was more complex both because of special restrictions, and because of the fact that four trucks were assigned to a network of streets:

In the snow removal problem we have constraints on how far a truck is able to travel before it has to return for salt or gas. We have other constraints such as the amount of times a street must be plowed. Therefore, in the snow removal problem we have more constraints in how we cover a particular territory. In the algorithm we are following an optimal path, without having any constraints. This method simply involves finding the shortest route among streets for one entity. Whereas in the snow removal problem we have 4 entities finding a shortest path.

(Campbell 10/14/87)

The class discussion which followed this writing brought out the most crucial difference—that the shortest route from point A to point B does not necessarily traverse all streets, whereas the snow removal problem requires that all streets be covered at least once.

Eventually students began making the connections for themselves, even creating problems for their own analysis. For example, after a class discussion on the assignment problem (where n tasks are assigned to n workers in a way which minimizes cost (or time) of the entire job) the following out-of-class journal entry was made:

We did the assignment problem today, and in so doing, reviewed Flood's Algorithm . . . I wonder if it could be twisted to the snow plow problem. We are assigning each driver (the operator) to certain tasks (roads) at a cost of certain times. But the cost of doing each road is the same for each driver. I wonder how that works out: say we have 3 people and 3 tasks that correspond with the following table:

		Road	
Cost	1	2	3
	1	4	5	6
Driver 2	4	5	6
	3	4	5	6

Following the algorithm we subtract the smallest current cost from each column to get:

		Road	
Cost	1	2	3
	1	0	0	0
Driver 2	0	0	0
	3	0	0	0

This is optimal, however, there is not a unique zero. So I guess it just simply doesn't matter, and the cost, for this table is a minimum of 15 units. That makes sense anyway.

(Hart 11/2/87)

This journal entry was evidence that a student, without my prompting, took responsibility for forming a link between a current technique and the snow removal problem. Clearly, a highlight of the course for me was finding this use of the journal to ponder "What if . . . ?" The student resolved for herself that no matter which driver gets assigned to which road, the optimum is the same in this example. This was evidence of individual creative thinking that students were not ready to undertake earlier in the semester when we wrote about linear programming models.

The Prospectus

As described in the course syllabus in Appendix 1, the prospectus was to be written to update a hypothetical supervisor on the progress and the plans. Generally, their reports fell short of my expectations; they seemed to lack substance.

My responses to their initial descriptions asked for elaboration or clarification:

How do shifts of employees affect the problem? Do you not always have a crew available? Why are parked cars relevant to your problem? Surely they are a "problem" for drivers of snow plows. But it seems they are a "constant" factor in your model. Or do you plan to show how much faster snow could be removed if parking were restricted? You mention laying chemicals in the last paragraph. Isn't this a separate problem, or do you plan to integrate it with your problem? How?

(Mett, 10/1/87)

As the work progressed, my comments on the second prospectus reflected their development:

Your statement of the problem is nicely done. But I'm confused by your statement of the objective in the middle of the first page. Isn't the "distance covered" fixed? That is, aren't you required to cover the entire distance of (primary) roads? Perhaps you mean to maximize the distance covered within a fixed time? This sounds more like the efficiency you are looking for.

⋮

You should explain more about the algorithm for finding an Euler circuit for each truck. What is an Euler circuit, and how is it related to this problem? Does such an algorithm exist, or do you have to write it from scratch? Does such a circuit even exist? Has any research been done on this topic?

(Mett 11/1/87)

Student comments at the end of the semester indicated that the purpose of the prospectus was not clear, and that my responses to them contradicted the premise that this writing was intended for someone other than an instructor/evaluator. In retrospect, I see that the students were correct. My responses concerned their model for the problem, and certainly sounded like they came from a professor rather than from a less knowledgeable supervisor. In the future, I will show students a good model prospectus so that they know exactly what kind of writing is expected from them.

At the beginning, this kind of writing served as a time monitor, keeping the project on schedule. Students commented after writing the first prospectus that it helped them clarify and simplify their problem, it helped them recognize their accomplishments and face their future.

The prospectus helped in several ways. Defining the problem with its objective and all of the constraints made it easy to see that this problem must be simplified first to make it more manageable. Then after making the problem simple and solving it with few or no constraints—start adding constraints. It has also made me see that I'm not sure exactly what my next step is.   (Hicks, 9/30/87)

Another student wrote:

... it has gotten me thinking about the project as a whole, such as how immense it is looking right now ... I think we have to get those maps and write down the priority streets and start with them and work our way down to the smaller streets. Or look at the smaller streets as they are divided up by the primary streets and work up from there.   (Love 9/30/87)

During this preliminary stage the prospectus served the desired purpose of forcing students to encounter fundamental aspects of the problem, and to talk, work, and plan with each other. The first tendency was to gather a lot of information. Students interviewed city officials, measured roads, and revised city maps. Writing the prospectus helped them recognize that real progress would be made on the problem only after they were able to formulate their goals and simplify the problem. They eventually discovered that if they could solve a simple model, then they could gradually add variations which would make the model more realistic.

However, toward the end of the semester the monthly report lost its value. It became extra paper work, and did not help toward writing that final report, as I had hoped. Therefore I would recommend that a prospectus be used only until a working draft of the final report is developed.

The Final Analysis

Although the final results in the snow removal project fell short of my expectations, I must share the responsibility for any deficiencies. My review of the course has shown me that I had not clearly presented my complete agenda to the students. For example, students wrote their final report without a single reference to any literature on other problems of this type, even though I had "suggested" such a search in our individual conferences. Since I felt this aspect of their work was so important, I should have specified it as a requirement.

I had hoped for more sophisticated utilization of mathematical skills on their project. However, I was recently reminded by a colleague that realistic consulting problems are, in fact, often solved by the same persistence and common sense that these students demonstrated. It was difficult at times, watching them complicate a problem long before they ever tried to reduce and simplify, patiently letting them struggle to eventually arrive at a workable model.

Even though I had hoped for better searching and modeling skills, I was delighted by their final oral presentation. The students delivered a very professional set of results to a combination of members of the Radford snow removal staff and members of our mathematics/statistics department. I wrote in my own journal these impressions:

I was pleased we had an audience. They were interested, asked good questions, and made good suggestions. The presentation included one nice example showing how roads of equal length were adjusted to a new "length" which accounted for relative degrees of difficulty. I'm not sure Mr. Farlow will accept the recommended plan A, but both he and Mr. Chumley seemed open to persuasion—even if they didn't think we had a realistic impression of their difficult task. All students showed excellent poise in a well prepared presentation. The team work looked excellent. The response to questions was superb!   (Mett, 12/11/87)

Students had a variety of reactions to the oral presentation. One wrote:

... I think it went smoothly and I think we did a pretty good job of splitting the work up. I think it was a really good experience. I would have liked to have been able to add more to our paper afterwards, but I feel it did need to be written, for the most part, before the presentation.   (Hart 12/11/87)

A second student also compared the oral report with the written:

The idea of incorporating examples to help illustrate the ideas behind the project was more easily accomplished in an oral presentation than that of a written report. (Campbell 12/11/87)

Another student writes about an obstacle they had to handle during the presentation:

... The only problem which made me mad was that Herb Far-low had told us that the trucks could go anywhere from 5 to 25 mph. And then at the meeting he says that the trucks can't go over 5, or stretching it, 10 mph. I think he was a little mislead-ing with his first estimate.   (Hicks 12/11/87)

And a fourth student shows some personal impressions:

I was amazed, if that's the right word, on how many faculty members showed up. I hope that I did all right with my part of the speech. I was afraid that I screwed up part of it. I wondered if my drawings of the simplified network were of any help or if they were just a waste of time.   (Love 12/11/87)

Students felt that the project, including the written and oral reports, was valuable. Their only recommendation was a unanimous re-quest that an additional lab credit be added to the course. They felt that the two major reasons for this additional credit were

1. that the extra work involved with the project deserved more credit

2. that a scheduled lab would pre-arrange a common time for their group work.

In spite of any reservations about that formal writing task, the course project was a success. Where was a better place for students to have a first experience in groping? in working with a group? in tackling a problem where there are no clues on which chapter or which formula to use? in struggling with a situation where no best answer is certain? and in writing their way through difficulties and concerns and finally into understanding?

REFERENCES

1. C. Mett, "Writing as a Learning Device in Calculus," *Mathematics Teacher*, 80 (1987): 534–537.

Appendix 1

COURSE SYLLABUS
OPERATIONS RESEARCH
Statistics 441
Fall, 1987

TEXT:
Introduction to Operations Research
Hillier & Lieberman, (4th ed.)

INSTRUCTOR:
C. Mett
309A Reed Hall
Phone: 831-5026

OFFICE HOURS:
11–12:30	T, Th
1–2	T, Th
3–5	W

GRADING: Your grade will be based upon the following:
Project	30%
Homework	20%
Test 1	10%
Test 2	10%
Final Exam	20%
Journal	10%

The following accomplishments will guarantee the corresponding grade:
90–100%	A
80–89%	B
70–79%	C
60–69%	D

The final exam serves as a tie breaker in any borderline cases.

JOURNALS:
At least one entry should appear corresponding to each class pe-riod. This entry should contain:

1. A summary and discussion of what you learned in class in your own words.

2. Your personal accomplishments, reactions to exercises. Re-flection on how material can be useful to you and a discussion of how this material relates to other concepts you have learned in this or in other classes.

3. Any open questions and/or analysis of your difficulties and concerns.

Bring your journal to each class so that it can be used during class for thinking and discussion. Keep your writing in a small three-ring notebook so that parts of your journal can be collected on short notice. This writing will be credited for completeness only.

PROJECT:
Your semester will focus on a long-term project of your choice. Ideally, you will find a problem of interest to a local business or industry and will work with them as well as with your instructor. However, you are also free to choose a problem of personal in-terest. Whatever your topic, you must justify that it is a problem worthy of a semester's devotion. For example, consulting with a school system to plan its bus routes can be a problem which de-serves your time and energy, whereas finding the optimum route from your apartment to a parking place on campus would not be adequate to occupy a semester's work.

The project will be a group effort, and will be partitioned into the following parts:

| 30% | prospectus reports (details below) |
| 70% | final report (50% on results, 50% on justification) |

A tentative schedule for the project is given on the enclosed calen-dar.

MONTHLY PROSPECTUS:
The monthly prospectus consists of a brief 2 or 3 page report to a supervisor which contains:

■ *background*—a short reminder of the purpose of your project and its goals;

■ *accomplishments*—a detailed summary of what you have achieved so far towards your goals;

■ *projection*—a concrete set of plans on how you are going to spend resources (time, energy, money, etc.) during the next month.

The prospectus should be typed and written in a style that you would be proud to hand to the person who evaluates you for raises and promotions. The grade and credit for your prospectus will be part of the total project grade.

HONOR POLICY:

You are encouraged to work together on homework and projects
provided there is an EXCHANGE of ideas. You are expected to
acknowledge all assistance in the form of references (on written
work) or joint papers (on homework assignments). However, during
tests you are on your honor to neither give nor receive help.

Appendix 2

PROJECT SCHEDULE
OPERATIONS RESEARCH
Fall, 1987

Friday, September 11

Report choice of topic with a brief outline of goals, justification
of the need for work on this problem

Monday, September 28

First prospectus due: plan of analysis, anticipated results,
schedule, resources, allocation of labor/time

Monday, October 26

Second prospectus due: initial data and preliminary results

Friday, October 30

First draft due: rough discussion of preliminary results, interpretation, justification, and analysis (details in appendix)

Monday, November 2

Consulting on first draft

Monday, November 16

Second draft due

Wednesday, November 18

Consulting on second draft

Monday, November 30

FINAL RESULTS DUE, final prospectus due: wrapup of research and preliminary report outline

Monday, December 7

Final report due

Wednesday, December 9

Presentation to public (department and fellow students)

A Writing Fellows Program Meets an Abstract Algebra Class: The Instructor's and the Fellow's Perspectives

John O. Kiltinen and Lisa M. Mansfield
Northern Michigan University, Marquette, Michigan

This paper addresses the role of writing in the learning of abstract algebra, and relates our experience at Northern Michigan University using a "Writing Fellows Program" in a junior-level abstact algebra course. The instructor's perspective is given first, followed by that of the writing fellow, who was a student with a double major in mathematics and English. As an appendix, we have included a "writing guidelines" document which was prepared for use by the students as part of the project.

The Instructor's Perspective

An abstract algebra course is often the point in a mathematics student's career that he or she has a first serious encounter with abstract reasoning and rigorous proof. It is not an easy encounter for many. The ancient wisdom of Euclid's words to Ptolemy that "there is no royal road to geometry" applies equally well to abstract algebra. One does not master the subject without considerable effort.

Much of this effort requires skillful use of language. Some mathematical ideas can best be grasped by the mind in terms of images, pictures, graphs or formulas, but to understand the logic of a proof, one needs to express it in words. Mere memorization does not suffice, for one needs not only the words but the concepts which lie behind them. One can have the words without the ideas, but it is very difficult to conceive of having the ideas without the words. To obtain ownership of the ideas, one must have processed them for oneself through the medium of words. An important way to do this is through the writing of proofs.

To obtain ownership of the ideas, one must have processed them for oneself through the medium of words. An important way to do this is through the writing of proofs.

These ideas about the relationship of language to mathematics have guided the first author's teaching of the introductory abstract algebra class at Northern Michigan University for the past eighteen years. Students in the class are all required to do considerable writing of proofs. The expectation is that the students will give careful attention to the quality of their exposition as well as to the soundness of their reasoning and the accuracy of their computations.

The challenge of doing such writing is a great one for many of the students. Not only are they being asked to expand their ability for rigorous reasoning beyond what has previously been required of them, but at the same time they are being asked to achieve a clarity of writing that has possibly not been needed earlier in their studies. Students are being asked to integrate several high-level skills to accomplish a complex objective.

Moreover, any expedients which seek to break the process down into smaller steps for them are inimical to the ultimate goal—understanding the process of rigorous reasoning as an integrated whole. By the time one reaches the junior-level abstract algebra course, one must begin this integration process—drawing together the mathematical manipulative, the reasoning, and the language skills which ideally have been honed in earlier courses. (The writing skills in the context of mathematics are perhaps the most neglected among these, and we are exploring ways of building more writing experience into earlier mathematics courses at Northern Michigan University.)

To learn how proofs work, one must write proofs. To this end, I regularly give my abstract algebra students at least six assignments of written solutions to sets of exercises. These are generally ones taken from the exercise sets in the text, and are those of moderate difficulty. They generally have a week to ten days to do perhaps five such problems. When they turn in their assignments, I always give them a set of solutions which I have written up so that they get a series of examples of what I consider to be good exposition for the types of exercises they are working on.

The students are encouraged to work together with fellow class members and use outside references, but are expected to acknowledge all sources. (It is usually quite transparent who is working together with whom even if they fail to acknowledge it, and usually one gets a feeling for who is the source of the ideas in the working groups which form.)

Forty percent of the students' grades is determined by their performance on these problem sets, and they are made well aware that the quality of their writing as well as the accuracy of their mathematics will affect their grade. Another forty percent of their grade is determined by a mid-semester examination and a final examination. The remaining twenty percent is on the basis of "participation." This includes regular class attendance (which earns one a C for this category), participation in class discussions (good work here will move one up to a B) and work on certain "challenge problems" which arise from time to time. Some good work on challenge problems is necessary for a participation grade of A.

An opportunity arose during the 1988–89 academic year to experiment with a way of helping the abstract algebra students with their writing. A Writing Fellows Program was established to provide support for writing in a variety of courses throughout the curriculum. I eagerly applied to involve my abstract algebra class, MA 312, in the project.

The general plan was to assign writing fellows to courses without particular regard to their familiarity with the subject matter. However, I had talked about the program with Lisa Mansfield, one of the top students in my previous MA 312 section who is a double major in mathematics and English. She was interested in applying to work as a writing fellow. I specified in my application that I wanted to participate only if Ms. Mansfield were assigned to my class. I pointed out that the highly technical nature of the writing would make it almost impossible for someone not familiar with the mathematics to be of much help to the students. The committee granted this request, and Ms. Mansfield was assigned to work with my students. She worked with the class both semesters.

Each semester, she came to an early class session to meet the students. While she was there, we explained to them that they would be working with her on two of their problem assignments. They would first submit drafts of these to her for her commentary. After she had read their papers, she would meet with them individually to give her comments. After that, they would write a final version to submit to the instructor. During Ms. Mansfield's visit, we set up a schedule of appointments for the students' meetings with her.

In anticipation of her work with the students, I wrote a set of writing guidelines for the students. These put together some observations about style and form that I had been communicating orally with students for several years. However, with another person involved in the reading of their work, I felt it important to have these guidelines in writing for ease of reference.

After having written the document and having put it into use, I wondered why I had not done it years before. It is helpful to have a carefully articulated statement of one's expectations. Involvement in the program provided the motivation to get this done. The writing guidelines are included as an appendix to this paper.

The system worked quite well. Ms. Mansfield met with the students at least twice during the course. They benefited from her careful reading of their work and from her suggestions.

It is my perception that the students felt more comfortable with the writing assignments and more confident about their ability to handle them as a result of the program. A review of grades in my abstract algebra sections over the past several years reveals no major difference in the average grades received by the students with whom the writing fellow has worked. While we have not found the "royal road," we feel that we have a means of making the journey a bit less bumpy.

The Writing Fellow's Perspective

Words. Words are the tools by which each of us express ourselves every day. Our words say who we are and what we know. Language, the integrated working of words, can be persuasive, informative, and powerful. Throughout history, words have moved men to action and maintained inaction.

Recently, there has been a resurgence of effort throughout the nation towards the promotion of those basic skills which form the core of all academic programs. Writing is one, and perhaps the most necessary, of these skills. No matter what the employment context, at some point an individual will have to explain to another what he has done and what he has learned in the process. He will have to articulate this through oral or written discourse. For what good is any knowledge if it cannot be communicated to others who might utilize and expand upon it?

Writing is one of the primary components of communication. It is a lot more than putting grammatically correct sentences together. It is a precise and exact skill. Much as an artist develops his abilities, only beginning with a set of elementary rules, so a writer writes. As he strengthens his skills and his individual style, he learns when it is permissible to break the rules and how to gain effect from such abuse. A writer brings to his exposition his self, his knowledge, and his experience.

At some point, each of us has uttered or heard another utter the familiar phrase, "I understand, but I can't explain it." I have always been an advocate of the notion that one does not truly understand an idea until he is able to articulate it, but until recently, I had never applied this notion to mathematics. Yet when I was asked in an abstract algebra class at Northern Michigan University to do just that: to solve problems algebraically and then to verbally discuss the process, I discovered an integrated expansion of my knowledge that I had never fully employed before. Sure I had attempted to orally explain a mathematical process to others, and of course, I have symbolically written out zillions of mathematical equations, formulas, and general laws for the universe. This was something different!

I found myself understanding the nicks and crannies of the mathematical process as I never had before. I had to. I couldn't bluff my way through this kind of writing. Mathematical exposition is telltale: either the knowledge is there or it isn't.

I found myself understanding the nicks and crannies of the mathematical process as I never had before. I had to. I couldn't bluff my way through this kind of writing. Mathematical exposition is telltale: either the knowledge is there or it isn't.

In addition to clinching my mathematical knowledge, I found my writing sharpening. It had to be exact. Fuzzy ideas were hung out to dry. This exposition had to be clear and sensitive to its readers' level of comprehension. It must be straightforward and logical—much like mathematics itself.

Learning to write and write well takes practice. There is no right and wrong way. To learn, an individual must be willing to sweat and grow—to turn his thoughts inside out and stretch his imagination to limitless bounds. There are no "magic formulas" to successful writing. At most, there are common strategies which work for many. This, the promotion of writing as an expression of an individual's knowledge and personality, is the nucleus of what the new Writing Fellows Program advocates. The program, introduced and endorsed at Northern Michigan University (NMU) by the University Writing Committee, is an attempt to promote writing across the curriculum.

This new, experimental program is modeled after a similar one at Brown University. It employs upper-level students who themselves have been recognized by the faculty to possess superior writing skills. Each of these writing fellows is assigned to a class of approximately twenty students with whom he is intimately involved throughout the semester. After preparing a workable schedule with the course instructor, the fellow is then to read and make written comments on two papers written by each student. These comments may focus on grammatical, organizational, and even stylistic errors. The hope is that the peer evaluator can assist the student in locating his problem areas and eliminating them. For each paper, the student author is expected to meet with the fellow to discuss his strengths and weaknesses and what he may do to improve his revision. Often students feel less intimidated when scholastic advice comes from a peer—a student subject to many of the same pressures as themselves. Of course, the author is then left to accept or ignore the tutor's suggestions.

Finally, both the original and the revised versions of the paper are submitted to the instructor in order to insure the professor that the peer tutor is not misleading the students.

For the past two semesters, I have served as a writing fellow for MA 312, Abstract Algebra, the same course which stimulated my interest in mathematical exposition. My experience has been beneficial to both myself and the students I have come into contact with. If nothing else, math majors at NMU have become conscious of their writing—something most of them had previously ignored. Their consciousness alone has sharpened their expository skills. It became evident to many of them that a lot more thought transpires when solving problems than that which is symbolically indicated. Axioms, theorems, and logical jumps seem obvious while one is working a problem, but are often fuzzy when reviewing one's work. The written work accompanying formulas and algebraic manipulations supplies the missing links and records them for future reference. The students in abstract algebra have gained new insights. They proofread their math papers. When I receive a set of rough drafts, it is apparent who knows what they're talking about and who doesn't. A lot of symbols on a piece of paper can look convincing—a lot of words can't. The words reveal themselves.

Overall, the program is working: it is working well. Northern's Mathematics and Computer Science Department is presently investigating how this entire notion of mathematical exposition as a measurement of student comprehension might be incorporated in courses from prealgebra on up. Students are gaining practical knowledge outside the classroom. A new stimulus has been added to the organism of Northern Michigan University.

A lot of symbols on a piece of paper can look convincing—a lot of words can't. The words reveal themselves.

Appendix:

MA 312 Abstract Algebra I:
Guidelines for Written Assignments

You will be required to turn in at least six sets of solutions to problems from the text during this course. Part of your grade on these assignments will be based upon the quality of your writing. These notes are intended to give you some guidance on important matters of style of mathematical exposition.

1. A STATEMENT OF THE PROBLEM   You should begin your presentation of each problem with a carefully worded statement of what the problem is. *Do not simply repeat the problem the way it is stated in the text.* In the text, problems are usually stated in the imperative mode, using phrases such as "Show that . . ." or "Prove that . . .". You should rework this into a declarative statement of what it is that you will be proving. You may modify the statement to emphasize aspects which you feel need particular attention, or to reflect any generalizations of the original problem which you are making. A typical example of a textbook exercise is:

If f is a 1-1 mapping of S onto T, prove that f^{-1} is a 1-1 mapping of T onto S.

If you were writing a solution to this problem, you might begin as follows:

Exercise 3, p. 15: PROPOSITION. If f is an injective mapping from a set S onto another set T, then f^{-1} is also an injective mapping , which maps from T onto S.

Proof. If f is as hypothesized, we observe first that f^{-1} is . . .

Note that the abbreviation 1-1 has been replaced by the term "injective" which is the one which will be preferred. We have identified the exercise by its number and its page (this being from Herstein's book). We have also labeled our statement as a "proposition." If the statement to be proven were of broad enough significance, we would call it a "theorem," and if it were one that was not the main issue under discussion but needed in order to prove the main result, we might call it a "lemma."

The proof is clearly separated from the statement of the proposition by a blank line and the word "proof." A common error students make is to run the statement of the claim and its proof together so one cannot tell where the first ends and the second begins. Try to avoid this.

2. AUDIENCE   Every writer should have a clear concept of the intended audience for a piece of writing. Only then can the writer be sure that he or she is giving the right amount of information to effectively communicate. This is especially the case for mathematical writing. When mathematicians write for each other, they write in a very brief style with many details omitted. When they write for students, they include more detail. (Students, however, may feel that they have not included enough, and sometimes they are right.)

Every writer should have a clear concept of the intended audience for a piece of writing.

For the problem sets which you will be doing, you should take your audience to be your peers in the abstract algebra class. Write your solutions with the goal of communicating with one of your peers. You may of course assume that they are familiar with terms and theorems in the text, so you can cite it freely.

Since you are a member of the class, you may find it helpful to regard yourself as the audience. Indeed, writing of mathematical and logical ideas is an effective means of communicating with yourself. The discipline of putting the ideas down on paper in a logical sequence forces you to think them through clearly, and in the process you will find that you have convinced yourself of their validity.

3. THE MATHEMATICAL "CLEARLY"   You have no doubt noticed that mathematical writing makes frequent use of the term "clearly." This is a very useful devise for calling the reader's attention to some necessary details of the proof, and leaving it to the reader to think them through. The writer is usually omitting these details because to include them would distract from the main focus of the argument.

The use of "clearly" communicates to the reader that the writer regards these details as being easy to work out. Clearly this may not be the case for all readers. This is why it is important to keep the audience in mind on any given piece of writing. Also, the writer may be using "clearly" out of laziness, omitting details that should be included but which he or she has not taken the time to work out or organize in a presentable form. As you develop your sense of style regarding "clearly," you will want to keep in mind that your professor is also a member of your audience—a somewhat skeptical member. You may be able to get by with one of your peers with an unsubstantiated claim that "clearly" something is true. Your professor, however, is going to be skeptical of such claims until you have developed a track record of using this time-saving devise with skill. *Whenever you use the term "clearly" or something equivalent, be sure that you have checked the details of your unsubstantiated claim, and could supply them if asked to do so.*

4. GRAMMAR   It should go without saying that good grammar will be important in your writing. Pay attention to the structure of your sentences. The worst problem that MA 312 students have had over the years in this regard is the use of run-on sentences. Be sure that each of your sentences follows the generally accepted standards for English usage.

Writing about mathematics presents some unique grammatical issues, among them, the usage of mathematical symbols within the text of a sentence. It takes some skill and practice to integrate mathematical symbols into sentences in English and have them conform to the rules of grammar. Indeed, you can find many bad examples in mathematical textbooks. Many writers and editors are rather insensitive to this point, or have adapted a specialized set of rules for their own purposes.

To illustrate, consider the sentence:

For every $n \geq 2$, there is an $m \leq n$ such that $m|n$.

This sentence contains three mathematical expressions. Many writers of mathematics would find it quite acceptable. However, let us analyze it. The expression "$n \geq 2$" is usually read as "n is greater than or equal to 2." Similarly, "$m \leq n$" is read as "m is less than or equal to n" and "$m|n$" is read as "m divides n." Note that if you read this sentence with these verbalizations of the symbols, it comes out as "For every n is greater than or equal to 2, there is an m is less than or equal to n such that m divides n." Only the last of the symbolic expressions fits grammatically into the sentence.

Many writers of mathematics ignore the problem of how to verbalize the mathematical symbols in their sentences. They write sentences which compactly convey the mathematical information, and leave it to the inventiveness of the reader to verbalize them in a reasonable way. Most student readers will not have sufficient background to do this, and will either verbalize awkwardly or not at all.

We will strive for a better integration of symbols with words. Here is a reworking of the sentence above which reads more easily:

> For every integer n with $n \geq 2$, there is an m with $m \leq n$ such that $m|n$.

This version is just a bit longer, but it reads more easily. To verbalize it with good grammar requires the reader to read "$n \geq 2$" as "n (*being*) greater than or equal to 2" rather than "n *is* greater than or equal to 2." However, most readers will do this instinctively. If one wanted to make the sentence such that the standard reading of the symbols "$n \geq 2$" is possible, one could replace each of the occurrences of "with" by "such that" and replace the last "such that" by an "and." This makes for a sentence that is grammatically unassailable, but it comes out sounding a bit stilted and pedantic.

It will take practice to develop a good sense of integrating symbols into sentences. There are no simple formulas for what is the best approach in every instance. Our main concern is that you develop a sensitivity toward this aspect of mathematical writing, and a reasonable level of skill.

There is one situation in which it will be acceptable to use sentence fragments. That is in brief justifications for steps in manipulative displays. A section below addresses this issue in greater detail.

5. QUANTIFICATION   Quantification is one of the key concepts which separates abstract algebra from your earlier experiences with algebra. This term refers to the specification of the scope of variable symbols by means of the quantifying phrases "for all" and "there exists" or their linguistic equivalents. Whenever we introduce a variable symbol to represent some object in our domain of discourse, we always want to make it clear whether the symbol can be taken to be any of the objects in the domain, or if we are asserting by its introduction the existence of some object in the domain which has particular properties.

For example, the associativity axiom involves the equation $x(yz) = (xy)z$. An important feature of the axiom is that this equation is understood to be true *for all* of the elements x, y, and z under consideration, be they elements of an abstract group, a ring or the real numbers. On the other hand, the identity axiom for groups involves the equations $ex = x$ and $xe = x$. Here, the assertion is that there is a *particular* element e in the group such that *for every* element x in the group, these equations are true. These quantifying phrases are vitally important for the correct understanding of the mathematical ideas. You must be sensitive to using them well, reviewing your work to see that you have adequately quantified over variables.

Since quantifying phrases must be used so often, there is the danger that your writing can take on a dull, repetitive tone. To avoid this, you will want to develop a sense for the linguistic equivalents which are alternatives to "for all" and "there exists." For example, the word "whenever" combines universal quantification with the logical structure of implication. The sentence "$x^2 > 4$ whenever $x > 2$" is an efficient way of saying, "for any x, if $x > 2$, then $x^2 > 4$."

English gives great flexibility in placing the quantification within a sentence. For example, the following three sentences have the same meaning in spite of their different word order:

> For all m and n in the set of integers, if $mn = 0$, then $m = 0$ or $n = 0$.

> In the set of integers, if $mn = 0$ for any m and n, then $m = 0$ or $n = 0$.

> If $mn = 0$, then $m = 0$ or $n = 0$ for any integers m and n.

You will want to take advantage of variety such as this in order to keep your writing lively. You should also be aware of the conventional interpretation that variables which appear without quantification are understood to be universal. That is, if you read a statement such as "Since we have $ab = ba$, we know also that $(ab)^n = a^n b^n$," and there has been no quantification indicated for a and b, you are to understand that the claim is made for all a and b in the domain of discourse. One can make good use of this convention in situations where one wishes to remind the reader in passing of some universally true equation in a situation in which to add the quantifiers would draw attention away from the main point. The sample sentence above is a case in point. Here, the objective was to quickly remind the reader that the commutativity axiom has as a consequence that integer exponents distribute over products. To add quantification here would draw attention away from the next sentence, which presumedly states a consequence of this fact that is of interest.

6. SYMBOLIC MANIPULATIVE DISPLAYS   Any serious writing about algebra must make use at times of some extensive algebraic manipulations. It takes some attention and practice to develop skill at writing symbolic manipulations which effectively communicate with the reader. An important point to remember in this regard is that *the manipulative display which best communicates with the reader is probably not the first one you found when you proved the point for yourself*. That is, if what you write up is simply a chronology of your discovery process, your proof is most likely not going to be as clear to the reader as it would be if you spent some time refining your approach with an eye toward good exposition. For example, suppose you were proving that in a group G, if any two elements a and b commute, then a commutes with b^{-1} as well. That is, if $ab = ba$, then $ab^{-1} = b^{-1}a$. You may begin your proof by looking at the equation which is to be proven, and doing some manipulations with it. You might notice that if you multiply both sides on the right by b, then the following happens:

$$(ab^{-1})b = (b^{-1}a)b$$
$$a(b^{-1}b) = b^{-1}(ab)$$
$$a = b^{-1}(ab)$$

Then if you multiply each side of this last equation by b on the left, you get

$$ba = b(b^{-1}(ab))$$
$$ba = (bb^{-1})(ab)$$
$$ba = ab.$$

Thus, if the target equation is true, so is the hypothesis $ab = ba$. This is the converse of what you wanted to prove. You could simply reverse this development and get a valid proof.

One disadvantage of displays of the sort above is that the reader must study both sides of each equation to see what changes have been made and then consider why these changes are justified. This puts an unnecessary burden on the reader. A better approach is to take the materials which you have developed during your discovery process and rework them into what we will call a "one-liner" display. By this we mean a display which proves the equality of the two terms in question by means of a single string of equalities which begins with one of the terms and ends with the other. Such a display is easier for the reader to follow, because the focus of attention at each step is clear. The materials given above yield the following one-liner proof:

$$ab^{-1} = e(ab^{-1}) \qquad \text{Identity axiom}$$
$$= (b^{-1}b)(ab^{-1}) \qquad \text{Inverse axiom}$$
$$= b^{-1}(ba)b^{-1} \qquad \text{Gen. associative law}$$
$$= b^{-1}(ab)b^{-1} \qquad \text{Hypothesis: } ab = ba$$
$$= (b^{-1}a)(bb^{-1}) \qquad \text{Gen. associative law}$$
$$= (b^{-1}a)e \qquad \text{Inverse axiom}$$
$$= b^{-1}a. \qquad \text{Identity axiom}$$

This display allows the reader to focus at each step on a few easily found changes from the previous line. The brief explanation given at the right gives the justification for the step. (Note: This is the one circumstance in which sentence fragments will be acceptable.)

Another aspect of good usage of displays has to do with how they are introduced and how they are interpreted. It is not good usage to write down a series of symbolic manipulations without any introduction in English. The introduction must be there to address issues such as the quantification. Also, after the display, you want to give a sentence which states the significance of the manipulation for the question under consideration.

7. NEATNESS Your written work may be done by hand. (Indeed, when one is working with mathematical notation to the extent that we will be, it is unreasonable to even consider submitting typed work.) You should expect to work out your solutions on scratch paper. Only when you know where you are headed should you begin to write up the copy which you plan to submit.

It is highly recommended that you write with a pencil rather than with a pen. You can count on there being a need to make minor changes before turning the work in. It is better to be able to erase. Neat and elegant handwriting is not expected, but legibility is.

8. THE USE OF DIAGRAMS You have heard that it was said to the men of old ... "one picture is worth a thousand words." This is nowhere so true as in mathematical exposition. (Well, if not a thousand, then at least a few dozen.) You will want to learn to effectively use diagrams to help communicate your ideas.

Our textbook only gives a few examples. However, during the lectures and on the instructor's write-ups of the problems, you will be getting a number of examples. Here is one. A *homomorphism* from a group G to a group H is a function α such that for every a and b in G, $\alpha(ab) = \alpha(a)\alpha(b)$. The idea of this definition can be pictured with the diagram opposite:

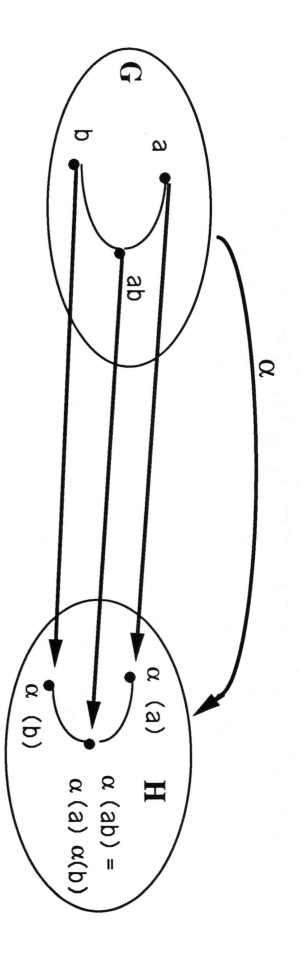

By itself, this diagram does not do an adequate job. However, with a few sentences of explanation and with the equation, it can help to visually convey the idea.

This is an important feature of diagrams: *They can often serve as good storers of insight and information.* That is, if you can remember a diagram which captures the essence of an idea, you can often reconstruct the details from it and thereby recover the whole idea. If you learn to use diagrams well in your writing, it will probably benefit you as much as or more than it benefits your reader.

9. "In Other Words"   This phrase is one which should occur often in your writing. You should always be on the lookout for ways of taking ideas which have been stated using equations and restate them using words. The reason is that words can convey the essence of a concept in a way that equations cannot, even as equations can convey the precise details very efficiently in a way that words cannot. To communicate most effectively, you should strive to use both tools.

We have already seen an example of this above. Near the end of Section 5, we gave an example involving the equation $(ab)^n = a^n b^n$. We stated the idea of this equation in words using the phrase "integer exponents distribute over products when the factors commute with each other." As another example, we can take the theorem "If $ab = ba$, then $a^m b^n = b^n a^m$," and state the idea of this in words by saying that "if two elements commute, then their integral powers also commute."

These statements of the ideas in words give yet another useful perspective on the mathematical realities. Stated in words, their ideas are often easier to remember. (How do you remember the sum formula for derivatives? Probably in the words "The derivative of the sum is the sum of the derivatives," not by means of the formula.) By using the phrase "in other words . . ." and restating your ideas, you will be helping your reader by putting the concepts in memorable form. You will probably also be helping yourself even more.

10. Acknowledging Sources   You are encouraged to work together and use outside resources as you prepare your written work. If you do so, common standards of academic honesty require that you acknowledge the sources of your ideas. *This should be done by means of a note at the end of each exercise.* These notes should be as explicit as possible about which insights you obtained from what sources.

11. Read It Aloud   A very helpful device to use to check if your writing is grammatically correct and reads smoothly is to read it aloud. When you hear what you have written, you often notice problems which slipped by during a silent reading. This can be especially helpful with regard to the integration of mathematical notations into sentences as discussed in Section 4.

A very helpful device to use to check if your writing is grammatically correct and reads smoothly is to read it aloud.

It is hoped that these notes are helpful to you as you begin a serious encounter with mathematical writing. The task is not an easy one, but the skills you develop will be of lasting value.

ON KEEPING JOURNALS

Writing Abstracts as a Means of Review

David G. Hartz
College of Wooster, Wooster, Ohio

In the spring of 1988 I was scheduled to teach "Finite Math for Social Science Students" as a follow up to the "Calculus for Social Science Students" course I taught the previous semester. This calculus course was a disaster. The typical student in the class was a junior or senior who was taking the course only because it was required for their major. These students were generally mathematically weak and this weakness was compounded by the fact that many had not taken any mathematics for several years. They did not like math, they did not understand math, and they resented the fact that they were forced to take math. Needless to say, they did very little work.

With this background, I was not looking forward to teaching the follow-up course to these same students. I needed to come up with some idea to encourage the students to work steadily throughout the semester. In the previous course the students had the tendency to let things slide until a day or two before a test and then they tried to cram a month or more of material into one evening study session. I was determined not to let this happen again. I wanted to make the course have more interest to the students while at the same time require the students to keep up with the material.

To accomplish these goals I assigned a weekly project: Write an abstract of the previous week's material. The point of these abstracts was to force the students to regularly read their notes and to think about what was covered. By taking the time each week to write a page or two on the week's work the students could not help but review and keep the material fresh in their minds. By relating the material covered to their own interests and disciplines, the students were able to appreciate the value and utility of mathematics. Other objectives of the abstracts assignments were to help to discover any mistakes and misunderstandings which the students may have had and to try to clear them up before any permanent damage was done, to give me feedback from the class, and to open up a line of communication with the class which might otherwise not exist. The most surprising part of this course requirement was that, in the end, the students thought the abstracts were a good idea. They found them to be useful, and, in some cases, even fun. I was encouraged to continue to use them in future courses. Because of this strong reaction, I have assigned abstracts in both my first- and third-semester calculus classes with similar success.

Assignment

I assigned the class to write a one, or at most, two page abstract of the material covered during the previous week and their reaction to it. What the students wanted to write about was left for them to decide; however, I expected at least some mention of each major topic covered during the week. I wanted the students to try to explain, in their own words, what we did. I stressed that the intended audience for the abstract was a student in the class. The students were not to write the abstracts as if they were explaining what we covered to me. I told them that I already knew the subject and didn't need it explained to me; rather they should explain it to someone who didn't understand the material and needed their help. The students must write in proper English using their own words, and mathematical formulas and equations were to be kept to a minimum. The students understand things better if they can explain it without resorting to the formulas.

I assigned the class to write a one, or at most, two page abstract of the material covered during the previous week and their reaction to it.

At the beginning of the semester I gave the students a list of questions they to consider: What are the major ideas behind the topic? What is this used for? Why are we doing it this way? Are there any ties that they recognize between what we covered in class and what they have studied in other classes or outside of school? What questions do they have? Was it easy? Was it difficult? I also asked the students to inform me how they thought the course was going. Was anything covered during the week which they didn't understand at all? I also asked them if they had any ideas or suggestions on how the course could improve.

The students were encouraged to spend 20 or 30 minutes thinking about what we covered during the previous week and what they wanted to say about it before they started to write. By reviewing what they had learned, they would have a better understanding of the concepts and this would make the abstract easier to write. Even more important for them, I told them that this would also improve their test performance. I have not done any testing to check this, but I have noticed that I have had fewer students fail the course in the classes in which I have used abstracts than in my other classes. In this way, I hoped to encourage the students to make a steady, regular review of the material and not leave things to the last minute.

Grading

The abstracts are intended as an aid to assist the student in understanding the material. The students had to struggle to express their understanding in their own words, and I did not want to penalize them if their explanation did not turn out right. Therefore, I did not grade the abstracts on the correctness of what they wrote, but I did correct their mistakes. If I had given them a grade for correctness, the students would have been encouraged to take the safe approach of just rehashing what I said in class or what the book said. Rather, I prefer the students to make an effort to explain the material on their own. In the struggle to explain mathematics in their own words, with a minimum of formulas and equations, the students are forced to make a first step at understanding. I corrected the grammar and spelling but I did not penalize for these mistakes. However, I did return a few papers which were written in outline form to be rewritten in complete sentences before they would be graded.

For the first two semesters in which I assigned abstracts, I only graded them on completeness. I assigned a grade from 0 to 5 based on how complete [in terms of major topics mentioned] each abstract was. Most abstracts received a grade of 5. The abstract grades were averaged at the end of the semester to be worth half of a test [50 points or about 9% of the total grade]. Eventually some students began to realize how simple my grading system was. I received several abstracts which were little more than lists of the major topics covered in the previous week. This was all that was required under my system to receive a 5. Therefore, the grading system has evolved so that I grade both completeness and effort. Now a listing of the major topics would only receive a grade of 2 points. In order to receive more points the students must put more effort into their work.

I do not indicate the points the student receives on each abstract. This would only emphasize the grade aspect of the assignment and detract from the true value of the assignment as a tool to help improve the student's understanding of the material. I do make extensive comments on the paper, especially early in the semester, explaining what the student did well and where and how they can improve. I also try to give detailed explanations of any mistakes the students make and answer any questions they ask. This has turned out to be a very good method of starting a dialogue with some of the quieter students in the class.

It is very important to return the abstracts to the class by the next class period. This way the students feel their concerns and problems are taken seriously and responded to immediately. I am also able to quickly go back and review material which the entire class didn't understand before too much new material is covered. A number of times I was forced to change my Wednesday class to a review of some concept which didn't go well the previous week—clearing things up early before the next test caused permanent damage to their grades.

This need for immediate grading is a major problem with assigning abstracts. Grading 30 or 60 abstracts every Monday is a time consuming task and a large block of time must be budgeted to accomplish this. A writing assignment such as this should not be assigned if the time cannot be arranged for prompt grading. Otherwise, the assignment will just become a drudgery to grade and will be of little benefit for the student.

Benefits

The abstracts help the students in many ways. Some students find it exciting to be able to relate the mathematics to their own disciplines. By doing this and realizing the importance of mathematics they become much more interested in the course. There was an economics major in my "Calculus for Social Science Students" class who was just an average student. He attended class regularly and did his homework but he wasn't one of my better students. However in the "Finite Math for Social Science Students" class he was transformed. He seemed to find the mathematics much more interesting because, while writing his abstracts, he was relating the work we did to economics. A sample from his abstract on the week covering matrix multiplication and inverses illustrates the enthusiasm he showed for the course.

> At the end of the week the lecture centered upon 1.7, (the end of chapter 1). The Leontief Input-Output Analysis is, in its entirety, a model incorporating societal aggregate demand and supply which allows for scheduling of industrial production based upon input demand figures. The key information necessary is the expected societal (external) demand and industrial production requirements (internal). In reality, the most difficult variable to determine would be societal demand which could fluctuate with income, taste, import levels, and in the field of energy such an unforeseeable factor as weather. Internal factor of production requirements would be fairly uniform given a present level of technology and would need adjusting with progression in the area. An example of a more abstract issue involving both external and internal demands would be scarcity of a natural resource (coal, oil, land, etc. . . .)

> The relation between this type of analysis and relevant matrix work is that it serves as an advanced application of using matrix inverses. Due to the fact that total output is not only a function

> of external demand but internal requirements as well, a repeating variable (the factors of production considered—which are also the object of total output to be calculated), needs to be isolated on one side of the equation . . .

> Being an economics major perhaps this topic excited me more than others—good stuff, more!

Every week he wrote detailed, enthusiastic abstracts relating the material from our course to his interests. The abstracts encouraged his regular examination of the relationship between mathematics and economics and helped him to develop the interest and enthusiasm which made him the top student in the class.

The abstracts also encourage the students to think about mathematics, where it came from, and where it is headed. An example of this is illustrated in the following paragraph from an abstract written for my Calculus 1 class.

> The derivatives of the trigonometric functions are rather amazing when one thinks about it. Of all possible outcomes, $Dx \sin x = \cos x$. Simply $\cos x$; $\underline{not}$ $(1/542)\cos x(1/\pi) \cdot 2x$. But simply $\cos x$. Is it just $\underline{luck}$ on the part of the mathematicians who derived trig and calculus? I assume trig was developed before calculus, why or how could the solution prove to be so simple? Luck.

Again the abstracts encouraged the student to be more curious about the material than they otherwise would have.

In the Calculus 3 course, a student in the class became interested in generalizations of the directional derivative. In the abstract for that week he wondered about "directional integrals" as an inverse of the directional derivative. He had anticipated the line integral, at least along straight lines. In the following week he discussed his study of directional derivatives along a path, for example, the directional derivative of $f(x, y)$ in the direction of the path $y = x^2$, and the problems he was facing. Then in the next abstract he wrote the following about his ideas of integrating along these paths:

> I am still convinced that there is a way to differentiate in any direction (not just a straight line) so now I believe there is also a way to integrate in any direction. This I believe is easy to do even though I don't know how to differentiate in any direction yet. If for example you were to use $d(x^2)$ instead of dx you could integrate x^4 in terms of x^2 : $\int (x^2)^2 d(x^2)$; by making a substitution $u = x^2$, the process would be the same as we already do it; $\int (x^2)^2 d(x^2) = \int u^2 du = (1/3)u^3 + C = (1/3)x^6 + C$. Of course this is all in theory!

Here I was able to help a quiet student who was performing his own research and investigations into a problem he was interested in in mathematics. Without the abstract this communication would not have been likely.

Another use of the abstract is to help students who miss class and want to check whether they understand what they missed. By thinking about what they read and writing an abstract of it in their own words, they are forced to come to grips with the material. The fact that the abstract will be read and commented on gives them the reassurance that their effort at understanding will be checked. A portion of an abstract written by a student who missed a week of class illustrates this use.

> Due to my extended 'vacation' from class, this week's abstract is relatively important. I feel that I understand the material I've read, and I did successfully complete the problems assigned,

but whether I fully understand the concept behind the problems will now become evident! The idea of using linear approximation is easier to understand, I think, if one thinks of it geometrically. If one takes a value of $F(x)$ and adds to that value delta x times the tangent line, one gets an approximation of the true value of $F(x+ \text{delta } x)$. The slope of the tangent line times delta x is simply an estimation of where the function will be in delta x spaces. This process will then presumably work better with functions that have nice slow sweeping curves, than one that is rather erratic.

I have used abstracts for three semesters in a variety of courses. I have found the students are initially uncomfortable writing about mathematics. This is why it is important to make detailed comments on the first several papers to give the students encouragement and help build their confidence. After the students get used to writing, they become better at it and begin to even enjoy it. I received many comments such as: "I think that the use of abstracts . . . in calculus is *very* helpful . . . Writing abstracts has helped me immensely; & I believe that it is very advantageous for the student in learning calculus." "I really liked doing the abstracts—it was a great way for me to let you know how much (or how little) I know about what we are studying." And "I found that going though the notes and the text were helpful and the rethinking of ideas very good. I would like to see other math professors use them [abstracts]."

Overall, the vast majority of students who had a comment on the value of the abstracts were in favor of them. The students who did not like them generally felt the problems assigned were enough and the abstracts were too much. Some of the students who would have done well in the class without the abstracts felt they should only be required for the students who are doing poorly. Other students felt they should not be required every week. Some students just didn't see any benefit coming from the requirement but rather, the writing was just another demand on their time. "These abstracts haven't helped me at all [in understanding the concepts]. I felt as though they were just another demand on my time, not a useful study tool."

The students are much more open to expressing their concerns and problems in writing than they are in person either during class or outside of it.

I have found the abstracts to be a great idea. The classes using them have gone better and the students seemed to show a greater interest in the course. The weekly writing forces them to review on a regular basis and helps to prevent the students from falling behind. The students are much more open to expressing their concerns and problems in writing than they are in person either during class or outside of it. I am able to hear from all the students in class, not just the few outgoing students who usually dominate the classroom discussion. I have even found the abstracts encourage the quieter students to become more involved in class. I find a larger percentage of students actively participate in classes with abstracts than without. Thus the abstracts give me a very good gauge on the progress of the class. I am able to determine what went well and what didn't. I know when to spend extra time going over sections that caused problems or will cause problems. I am in strong agreement with the student who wrote in her final abstract:

I think the idea of writing an abstract each week is good. I do think it helps me understand what the actual concepts of math are. Usually a theorem is just a collection of words and derivations are just a collection of symbols, but having to write and explain what the equations actually mean helps me understand them instead of just plugging in numbers and finding the answer . . . Overall, I think abstracts are beneficial and once I got used to writing them, I couldn't stop!

Journals and Essay Examinations in Undergraduate Mathematics

Gary L. Britton

University of Wisconsin Center, West Bend, Wisconsin

Introduction

It has long been my opinion that undergraduate students should be expected to write about the mathematics that they are learning. There are two reasons for this.

1. If students are to thoroughly learn the material, then they should be able to write an explanation of the concepts, procedures, and key results of that material in order to demonstrate what they have learned.

2. The act of writing, as well as the study and preparation necessary to do that writing, are activities which aid the students' learning. It should help strengthen their mastery of the material.

For several years I have tried various techniques for incorporating writing into my calculus, precalculus, and statistics classes. Recently I have concentrated on just two specific ways to approach this writing expectation. These two activities are student journals used for weekly writing, and test questions requiring essay responses on examinations. This paper will discuss these two writing aspects of my courses.

Student Journals

One form of student writing which I have found useful is a weekly student journal. I will describe what I do and how I handle the mechanics of this aspect of a course, and then will discuss the advantages of such an activity.

JOURNAL PROCEDURES   Once each week students are expected to submit a journal which includes one to two pages of written material. So that the format is the same for all students every week, and so that the journal material for each student is together week after week, I have students use examination booklets. These are the familiar 24 page "blue books." They are issued to the students at the beginning of the semester and again when a new one is needed after the first book is filled. For students who prefer to use a word processor for writing, I allow them to submit the corresponding printout in lieu of a blue book. Students submit their journals at the beginning of the class period on the first class meeting of each week. I read the journals and usually return them the next class meeting. If I can't get them back by the next class, I make certain that I return them by the end of the week so the students will have them for their entries prior to coming to class the following week. If there is an exam scheduled on a day that the journals are due, students don't need to submit a journal that week.

I expect the journal entries to consist of any or all of a number of different types of material. The primary material is to be a summary explanation of the mathematics studied during the past week. I tell the students that if they do a good job on their journals, then an excellent general review before each exam would be to read their journal entries. Frequently the journals will include exercises that students have worked and want me to check for correctness

or else to determine why they can't get the correct result. Other journal material includes questions addressed to me, lists of items not understood, and comments about the course. I suggest to students that they might find it easier to write if they write the entries as diary entries or as letters to me. Some of them actually start out each entry with "Dear Diary" or "Dear Dr. Britton."

When I read the journals I try to write answers to all the questions. If the answers are too long, then I make a note asking the student to see me after class or during office hours. I also make written comments to try to correct any errors or misconceptions that appear in the entries. This approach is quite different than that suggested by Mett [3]. She prefers to never write in the student's journal as a matter of respect for their personal nature. I have never sensed any objection to the responses I write in the journals.

Students frequently list the problem numbers of text exercises that they could not do. I record these numbers in my own notes and then when I go to class I work some of those that were listed most frequently in the journal entries. This is a good supplement to, and sometimes replacement for, calling for exercises to work during class.

JOURNAL EVALUATION   I evaluate the entries and record those evaluations in my grade book. It is a rather subjective evaluation and I use only three levels. A minus if I think the entry is less than what I expect of most students, a check if it is satisfactory, and a plus for an exceptionally good entry. I do not accept late journals. To do so encourages irresponsible behavior and disrupts the schedule of returning journals by the end of the week. Most students submit their journals very regularly. For example in a recent trigonometry class of 18 students there were 11 journal entries expected during the semester. Of the 198 possible entries for all students, all but 22 were submitted. For those submitted I gave 60 pluses, 6 minuses, and 110 checks. The students' grades on the journals are then incorporated into what I call a course participation component of the final grade. It is usually around 8% of the final grade, which is similar to the weight given by Mett [3], though she determines the grade differently. Student evaluations of instruction indicate that some students think that the journals should not be used at all in the final grade, others believe that I should weigh them more heavily than I do.

ADVANTAGES   The primary reason for having students keep the journals is that they learn better if they need to describe the material they are studying. However, there are also a number of other advantages of the required journals. Three of these advantages are described here.

The increased communication between students and me is a definite advantage.

First of all, students must make an effort to keep up with their work. If they don't, it is quite evident when reading the journal entries. Of course some students will try to get by with entries such as "This week we studied section 4.5 on differentials and started section 4.6. This material was very hard." and it may take some effort on my part to get a student to make a more useful entry. But it appears that students do keep up with the material more now than when I wasn't using the journals.

Second, the increased communication between students and me is a definite advantage. Even though this communication is written rather than oral, it is communication. In fact I have had some students who have asked things or made comments in the journal that I know they would not have asked if they would have had to talk to me directly.

And third, one of the indirect advantages of using the journals as I do is that it is an excellent aid in taking attendance. Currently there is an increased interest among professors regarding the effect that class attendance has on the learning of course material. See, for example, Brown [1] and Cope [2]. I share that concern. On the day that I collect the journals, and again the day that I return them, I have a means of recording attendance without having to take roll during class.

STUDENT OPINIONS   How do students respond to the expectation of keeping a weekly journal? There are some who seem to think that it is a useless extra burden on them. Some think that since it is mathematics they shouldn't have to write. Many others think that the journals are quite helpful. In a free response form for student evaluation of instruction at the end of the semester, students seldom make negative comments about the journals. On the other hand, positive comments do appear. For example, these three were written by students in one class. "I believe the diaries are a good thing, but I believe that there should be no grade for these. Especially one that affects our final grade." "I particularly like his weekly journal requirement. The journal is effective in reviewing our lecture material." "The weekly journals and quizzes helped a lot because it kept you up to date on assignments."

It was his opinion that his journal would be more helpful to him than his trigonometry text itself.

In one class I had collected all of the journals at the end of the semester in order to try to do some analysis of them. One of the students from that class came to me during the following semester and asked if he could have his journal. He was taking a calculus class and wanted his journal in order to review some of the trigonometry. It was his opinion that his journal would be more helpful to him than his trigonometry text itself.

An algebra student wrote in the last journal entry for the semester "I really hated doing these journals. They were very time consuming considering no real grade was earned. It was a pain trying to remember to fill it out. However, I must admit that they were helpful in review and caused me to do some digging in the textbook and my brain. On a scale of 1–10 (10 being the highest) I'd give it a score of 8."

During a fall semester course with new freshmen I noticed that it took a few weeks for students to start making entries which were of the quality that I expected. Then I realized that they had no way of initially knowing what my expectations were nor what type of entries would be most helpful to them. So at the beginning of the following semester when I announced the requirement of the weekly journals, I also gave students a photocopy of one of the better entries from the previous semester. This gave them an idea of not only what I expected, but an example of what one of their fellow students had done. I found that as a result of this, the initial entries were much better than they had been the semester before.

Even though I generally don't need to spend more than one or two minutes reading each journal, it would be quite time consuming to do this for a large number of students. I have been using the journals in one or two classes of 20–30 students each, or else one larger class, rather than trying to use them in every class every semester. The results which I have found, even though evaluated subjectively, are positive enough for me to continue with essentially the same format.

Essay Questions on Examinations

Most college freshmen are surprised to find an exercise on a mathematics examination which requires an essay response. As examples of this type of exercise consider the two which follow. "Discuss the procedure for using the discriminant of a quadratic equation for determining the nature of the roots of the equation." "Explain what we mean by the inverse of a function." Even if students are informed, in advance of the exam, that they will be expected to write an essay response to at least one exercise such as these, they don't know how to study for it nor how to answer it during the exam. Students may respond by listing a number of properties or facts. They may respond by doing an example. They may respond by talking about the topic, but never indicating any real understanding of it.

REASONS FOR DIFFICULTIES IN WRITING   These difficulties are caused by the following factors.

1. Most students have not been expected to do a lot of writing in the past. This is true for all courses in general and is particularly true for mathematics and physical science courses.

2. Students have not seen an example of an appropriate response to such an essay exercise in mathematics. Consequently they have nothing on which to pattern their own work.

3. Students don't know for whom the response is being written. They assume that it is being written for the professor since that is who wrote the exam. Since the professor knows the material the student believes that thorough explanations are not necessary. By writing to the professor the student also assumes that the reader already knows the meaning of all of the terminology that may be used, and thus fails to give any explanation of terms.

Suppose that a friend of yours, who is also taking this calculus class, was sick and missed class the day we discussed graphing techniques. Write a letter to your friend . . .

SOLUTIONS   While we can't do anything about the first of these three factors, we can do something to overcome the problems indicated by the second and third. The simplest way to handle the second is to give the students a sample exercise and a sample response to it a few days prior to the first examination. On the first examination in a trigonometry course I had asked students to explain the six trigonometric functions using the unit circle as the basis for the definitions. Even though we had studied this and discussed it extensively in class, I was quite disappointed in the responses which the students gave on the examination. During the following semester I wanted to ask a question using that same format but decided to give the students a sample of the type of response which I expected. During the week prior to the exam I gave them a copy of the first semester exam and a one page essay which I wrote as a sample response to the exercise which asked them to explain the trig functions. While I don't have any quantitative evidence to show that the responses during the second semester were better as a result of this, my own observations of the writing, and the mathematical explanations of the trig functions, led me to conclude that there was a definite improvement in the quality of the responses. This is an example of how easy it is for us, as instructors, to give students some specific guidance in order to help them improve their writing.

We can eliminate the problems discussed in the third factor listed above by stating, for the students, the context in which they are to write their response. By this I mean that we can, and should, tell the students to whom they are writing, and how much mathematics they should assume that person knows. This helps the students decide which terms need explanations and which ones can be assumed. Perhaps some specific exercises, stated in the form in which I would use them on an exam, would be the best way to illustrate what I mean.

Sample exam exercise 1. Suppose that a friend of yours is taking trigonometry at another university. They have not yet studied the unit circle approach to the trigonometric functions. Write a letter to your friend explaining how the unit circle is used in defining the six trigonometric functions when this approach is used. You may assume that your friend has already learned the right triangle definitions of the trig functions.

Sample exam exercise 2. Suppose that a friend of yours, who is also taking this calculus class, was sick and missed class the day we discussed graphing techniques. Write a letter to your friend telling her about the use of derivatives in graphing a function $y = f(x)$. You may use an example to illustrate the procedures but be sure to include a thorough explanation using complete sentences and appropriate grammar.

Both of these let the students know that they are to write their response for another student, not for a mathematics professor! That is, they are writing for someone who knows some of the relevant mathematics, but doesn't know anything about the specific material which is to be explained. Hence, the explanation needs to be thorough, clear, and meaningful for the reader who is seeing this particular mathematical concept or procedure for the first time.

I tell students that after writing their response they should read what they have written, pretending they don't know anything about the topic being discussed. They should then ask themselves "After reading this response, would I know what I need to in order to understand the material being discussed?" For the second example above, that means that the students should pretend they don't know anything about using the derivative to help graph a function. Then after reading their own explanation they should decide if that explanation provided enough information in order for them to use the derivative to accurately draw the graph of a function.

Again, I don't have hard evidence of the effectiveness of posing test exercises in this manner, but the student essay responses that I get are more complete and more appropriately written than they were before I started using this method. I have shared the idea with English department colleagues who teach writing and they think that the approach which I have taken is a very useful one.

Conclusion

Students should be expected to do some writing in their mathematics classes. The instructions and expectations must be clearly stated and guidance must be given regarding the type of writing that one expects in mathematics. Examples, either prepared by the professor or written by a former student, should be distributed to current students. When carefully selected procedures are used, most students at my institution will not object to this expectation of writing and many will even admit to finding it helpful. Areas of future investigation include the following questions.

1. Do writing activities such as those described above increase students' mathematical retention and their abilities in problem solving?

2. Does grading journals influence student writing in those journals?

3. What are the cumulative effects of prolonged writing expectations? (e.g., extensive writing used during each of the first four math courses)

4. Does the use of writing activities affect mathematics appreciation?

REFERENCES

1. William R. Brown, "Point of View," *Chronicle of Higher Education*, 33 (January 28, 1987).

2. Charles L. Cope, "The Effect of Absences on Course Grades," *AMATYC Review*, 7 (February 1986): 37–41.

3. Coreen L. Mett, "Writing as a Learning Device in Calculus," *Mathematics Teacher*, 80 (1987): 534–537.

Weekly Journal Entries—
An Effective Tool for
Teaching Mathematics

Louis A. Talman
Metropolitan State College of Denver, Denver, Colorado

Introduction

In his remarkable and beautiful little monograph entitled *Liberal Education*, Mark Van Doren [2] wrote "'Language and mathematics are the mother tongues of our rational selves'—that is of the human race . . ." We, teachers and practitioners of mathematics, are well aware of the capacity of mathematics to empower the human mind; but I believe that we often forget or ignore the potential power of language, especially written language, as a tool for investigation. Van Doren reminds us that language is coequal with mathematics in its capacity to enable our minds. It is the purpose of this note to discuss an application of written language to the task of teaching (and learning) mathematics.

In *Everybody Counts: A Report to the Nation on the Future of Mathematics Education* [1], we read "Research on learning shows that most students cannot learn mathematics effectively by only listening and imitating . . . [and] . . . that students actually construct their own understanding based on new experiences that enlarge the intellectual framework in which ideas can be created." Let us first observe that writing about mathematics forces construction of understanding, because we cannot write coherently about something we do not understand. Professors of English tell us again and again of the deep interplay between language and thought.

Writing about mathematics forces construction of understanding, because we cannot write coherently about something we do not understand.

It occurred to me that one way to harness that interplay to the task of "enlarging the intellectual framework" within which students of mathematics work is to require them to keep journals which they must periodically submit for critical evaluation. I have subsequently experimented with the idea, and I will describe the results of my experimentation in what follows.

Methods

In several classes over the last year and a half, I have required students to submit weekly journal entries. These courses have so far ranged from a survey course for liberal arts students, through beginning and intermediate algebra, and into first semester calculus. I have not yet used the technique in an upper division undergraduate course, but I foresee the possibility—indeed, the likelihood.

In the courses where I have used the technique, I have begun the semester by handing out a sheet, entitled "Your Journal," at the first class meeting. The handout describes in some detail what I expect them to do with their journals. As I refine my ideas, this handout continues to evolve. The appendix to this note is the version I used last; the version I use next will be somewhat different.

In the assignment, I have instructed students to prepare and submit weekly journal entries. I, and they, have found it most convenient to have the entries due on the first class meeting of each week.

Each entry thus dealt with the work of the previous week. A weekly entry could be organized as the student wished, either as a single essay dealing with the week as a whole, or as a collection of smaller essays that each deal with shorter periods of time. Regardless of organization, the entry was to contain three kinds of writing: a short summary of the topics we covered during the week, a quick report on the student's own relevant activities for the week, and a lengthy analysis of the week's work. I have told students that their journal grades will depend on the quantity and quality of their reported activities; the content of the entry (especially the analysis); and grammar, spelling, and punctuation. I have not told them, although I plan to do so in the future, that their journal grades will *not* depend on the correctness of the mathematics they include.

I have recently required that each student analyze the solution to at least one problem that we did not solve in the classroom.

In fact, I have been principally interested in their analyses, and I have based the bulk of their grades on them. I have not really graded the quantity or quality of their activities, nor have I graded grammar, spelling, or punctuation—except in cases of flagrant inattention. I have mainly used the threat of doing so to keep them honest. I think this is especially important where grammar, etc., is concerned. They must pay attention to their grammar, spelling, and punctuation. These matters are part and parcel of the *habit of precision*—which is one of the things that I intend this writing assignment to help them develop.

I have recently required that each student analyze the solution to at least one problem that we did not solve in the classroom. In the absence of such a requirement, I have found that I see a good bit of stuff that is more than a little reminiscent of things I have put on the blackboard. That defeats another purpose of the assignment, which is to get them to attempt some mathematics on their own. As we all know, they find it much easier to regurgitate our thoughts than to have their own—and many of them *will* regurgitate, given the choice.

The first semester I used the journal technique—and every time thereafter—I included in their instructions a requirement that, in the entry that covers a week in which they sat for an exam, they devote a portion of their analyses to the exam and their performances on it. My intent was to get them to think about the exam after they had finished taking it—an activity most of them don't indulge in.

This requirement turned out to be an especially good one, owing to the fact that I habitually schedule exams for the last class meeting in the week. I do this routinely in order to get the entire weekend in which to do the grading, so this habit wasn't in the front of my mind when I included the requirement. But the happy result has been that I have required them to analyze their exams *before* they see *my* analyses of those exams. Mediocre students aren't very happy about this, but it is exactly what they need to do, and it is exactly what the better students have always done—though perhaps less formally than I now require.

An outgrowth of the requirement that they analyze their performance on an exam before I have returned their papers deserves discussion. On one occasion, one of my better students in a Calculus I course complained in her analysis that she couldn't recall what she had done on a particular problem. It turned out that the problem in question was one that she had botched—owing rather clearly to her failure to comprehend a central idea; she had, on the other hand, very satisfactorily recalled the correct work she had submitted for the other problems. In my comments on her entry, I suggested that perhaps the fact that she couldn't recall what she had submitted was a danger signal. I further suggested that she might be able to use her ability to recall her solutions to problems she had worked while studying as a measure of her understanding of underlying ideas. She agreed that the criterion had potential, and continued to experiment with it. In her final journal submission for the semester, she reported considerable success in using this criterion to identify matters that needed further attention.

It is important to note that grades on journal entries do not depend on the correctness of the mathematics included. *This is vital!* What matters is the effort that a student puts into thinking about mathematics, into thinking about thinking about mathematics, and into organizing those thoughts for presentation in writing. Any *bona fide* effort to analyze a problem: any effort that shows organization, willingness to experiment, effort to trap error; any effort that displays any of the things that you or I do as we do our mathematics deserves credit—whether the mathematics is correct or not. It is primarily their *approach* to mathematics and their *presentation* of the result that I grade.

Any *bona fide* effort to analyze a problem: any effort that shows organization, willingness to experiment, effort to trap error; any effort that displays any of the things that you or I do as we do our mathematics deserves credit—whether the mathematics is correct or not.

That is not to say that I let incorrect mathematics slip by uncorrected. I comment extensively on the mathematics they are writing about—as well as on the writing itself. At first, my policy of correcting their mathematics without penalizing puzzles them. Students expect to be punished for their mathematical mistakes. They are used to trying to *obfuscate* potential errors, or at least to trying to deny responsibility for them—often by using the passive voice. (Sample: "A mistake was made in calculating the derivative, so that the second critical point was not found." Note that no-one is responsible for this mistake—it just happened! We might ask ourselves who *taught* them to obfuscate, and after we recover from the ensuing discomfort, we might accept our own responsibility.)

This issue is the main problem that I must deal with during the first weeks of the semester. I must *retrain* students to attempt to *present* their thinking, potential errors and all. They must accept responsibility for their mistakes, when they would much rather conceal them and blame nobody in particular—least of all themselves. This retraining takes time, and they find that I get fussier and crankier about their evasions as the semester lengthens and they catch on. They *must* display their mistakes, because those mistakes are the high-grade ore I mine; they are what I use to teach.

At the end of the semester, I discard the lowest fifth of each student's scores and average the rest. I base one-third of the semester's grade on the resulting average. This is a substantial portion of their grade, and I expect a substantial effort from them for it. There are always a few who learn this the hard way. Every semester, the first entry includes several one- or two-paragraph submissions, hastily scrawled on a scrap of paper—even though the assignment clearly indicates that I expect several pages at least. Those submissions get what they deserve: A mark of zero and a note "This is not acceptable; please re-read the assignment." Second offenses are rare.

Advantages

This is an effective approach to learning mathematics. The connection between language and thought is real, and students *can* use it to help themselves "construct their own understanding".

> It never seems to fail that no matter how often I'll work a problem incorrectly, I'll invariably do it right when I go to prepare my journal. It really works to have just one problem in front of you, and to walk slowly through it and explain each step.
> —V.S., Calculus I

> In selecting problems for discussion in these entries I try to select ones that interest me and give me an opportunity to explain something to myself or puzzle out a seeming contradiction.
> —R.T., Calculus I

> At first I hated this journal idea and could not make it personal. After a while I began to see how it aided me in figuring out calculus and learned to use it as a means of putting my thoughts together in a useable form. It has helped me to test my level of understanding which has helped to enforce my skills and confidence. Although it is a lot of work . . . I feel that these journal entries are of great benefit.
> —H.E., Calculus I

Students repeatedly remark that they have just succeeded in solving, as they worked on their journal entries, a problem they didn't think they would be able to do when they sat down to write. Of course, I anticipated that this would happen, or I would not have made the assignment. But I did not anticipate the frequency with which it happens, and I do not tell students that it will happen because I think that the lesson is more effective when they discover it for themselves.

A second advantage: The journal entries open up a new and useful channel between student and instructor. I see firsthand how students are thinking; I see what misunderstandings they are laboring under; and I see what they really do comprehend. And having seen, I can intervene. I can head off developing misconceptions; I can dispel them while they are no more than nebulosities. Of course this depends very much on students' coming to realize that their journals are an environment where they can explore and make mistakes without paying a penalty. Once they have this understanding, I find that they will discuss almost anything.

> I look at [studying calculus] as a journey, and these entries were like taking an experienced guide along.
> —R.T., Calculus I

For many students, exploring mathematics with *any* kind of success is a new experience. They are amazed and pleased with themselves and their newly discovered capabilities.

Students cannot escape exploring mathematics in their journals. This is another advantage of the technique. Successful journal entries always involve personal exploration of some aspect of mathematics—however trivial. These explorations may fail in one of the narrow senses that the explorer gets lost or fails to find anything of real interest. But they are successful in the broader sense that the student becomes an explorer. For many students, exploring mathematics with *any* kind of success is a new experience. They are amazed and pleased with themselves and their newly discovered capabilities.

Still another advantage arises from the fact that students are writing about mathematics. Writing about mathematics is just a small step away from writing mathematics itself, and in fact, many students really do write mathematics for the first time. In working on their journals, all of them begin to learn how to write mathematics.

As a further bonus, I find that I get more out of my best students than I have in the past. I mean the ones who are capable of sailing through a lower division course unfazed. I have had the feeling many times in the past that I do not really give full measure to these people, and that they do not really give me full measure. The journal is a tool for changing that. Because the communication in the journal is one-on-one, I can teach these students more. And demand more of them in return for their A's.

I find that I get more out of my best students than I have in the past.

Finally, students like the technique. They gripe, but generally in the good-natured fashion of one who is working hard, knows it, is accomplishing something worthwhile, and knows that, too. I must admit that this advantage of using journals isn't of great importance to me, personally. I suspect that if I believed that I could instill understanding or appreciation of the Mean Value Theorem by smacking someone up alongside the head with a two-by-four, I would try that technique, too.

Disadvantages

The major disadvantage of using weekly journal entries has to do, as the reader has probably guessed, with grading them. I have perceived a two-fold disadvantage. First, grading weekly journal entries is a time-intensive chore. One must read them attentively—even more attentively than one reads exams. One must provide extensive commentary, especially during the first few weeks of the semester while one is trying to break their old habits and to instill new ones.

During the first semester I used journals, I went somewhat overboard. I used them in three courses simultaneously, and I think that I was reading somebody's journal at just about every odd minute of every day. (And during some of the even minutes, too!) I felt like I wasn't doing anything else! I learned from my mistake, and I now use the technique in just one course each semester. And I choose that course carefully as the one where I think the technique has the most to offer.

The second of the disadvantages associated with grading journal entries is perhaps more troubling to many of us. As mathematicians, we are likely to be uncomfortable with the idea of grading writing. We worry about our qualifications where writing is concerned, and to us the grading of writing seems more subjective than the grading of mathematics. These concerns are probably less justified than we think. We are, after all, highly educated folk, and we *can* **reliably** determine whether or not undergraduates are having reasonable ideas and whether or not they are expressing them reasonably well.

A final disadvantage: Students hate the technique. Not all of them, of course. By way of example: I announced on the first day of a calculus course that I would require them to hand in weekly journal entries during that semester, and I handed out the journal assignment sheet to 30 students. The following day, 18 students returned—I never saw the other 12 again. But then, I would probably have lost those 12 anyway, later on in the semester when both they and I had more substantial investments in failing enterprises. Come to think of it, maybe this one isn't a disadvantage after all . . .

Thoughts for the Future

I have no doubt that I will continue to use this technique. I find that the advantages clearly outweigh the disadvantages, and except for the drain on my time, the method fits my style. There are, however, some things that I will do differently.

Be explicit about what doesn't count toward the grade. In particular, I think that it will help me retrain them to present potential errors honestly and forthrightly if they know from the beginning that I will not punish them for their mathematical mistakes.

Provide more explicit instructions on what to do. There are several issues here. I need to encourage students to write about the problems they couldn't solve (that they are reluctant to do so is probably very closely tied to their fear of displaying their errors). I need to provide them with some instruction in dealing with seemingly intractable problems. Some of the material in *Thinking Mathematically* [3] may be helpful here, either directly or indirectly. In general, I need to give them some concrete strategies for conducting what I have in the past called "analysis". It may even be helpful to give them some examples of what other students have accomplished before they have to submit their own writing.

Involve the best students in self-evaluation, with their evaluations of their own work being the principle determinants of their grades. This is an entirely new idea, which occurred to me during the preparation of this note. The idea, which I shall have to think about some more, is to return the student's entry, ungraded, after several days, with instructions to the student to evaluate his own work for his grade.

APPENDIX

Your Journal

You are to keep a journal for this course. You must make **at least one** entry in your journal each week. At the beginning of the first class period of every week, you are to submit the entries you have made during the preceding week; late submissions **are not acceptable**. Journal entries should be submitted on ordinary 8 1/2 by 11 paper which is (or can be) punched for placement in a notebook.

A complete entry for a week will contain:

A. A *summary*, **written in your own words**, of the material we dealt with during the week for which you are making the entry. This summary need not be more than a few paragraphs long.

B. A *report* on the individual work you have done outside of class (problems you attempted—whether you solved them or not; reading you did—whether from the text, the handouts, or outside sources; participation in discussions—in the classroom cr out; or any other thing you may have done that you think relevant to the course). This discussion may be from one to several pages in length.

C. An *analysis* of the week's work and the understanding you have gained (or failed to gain) during the week. You may include: difficulties you may have experienced, extensions you may have thought of, questions you may still have, and successes you have enjoyed.

If you took an exam during the week preceding the entry, you should devote the bulk of this section to a thorough analysis of the problems on the exam, their solutions, and an evaluation of the solutions you submitted. In other weeks, you may choose to focus on a single problem or topic, or you may analyze the entire week's progress. **In any event, you must analyze the solution to at least one problem that we did not solve in class.**

You will find that a careful written analysis of the difficulties you encounter on problems you don't know how to solve is one of the most profitable things you can do in your journal.

Your analysis of your week's work is the heart of your journal; it should be at least several pages long.

I will grade your work on the quality and extent of your activities, the content of your remarks—especially in your analyses, and on your grammar, spelling, and organization. Within a reasonable time of your submission, I will return your work to you. You are then to undertake any rewriting I have asked you to do and place both the original entry and the rewritten entry into your journal notebook— which you should maintain separately from your classroom notes. I may ask you to submit these notebooks for inspection from time to time, so it is important that you keep them up to date and that you do the rewriting I ask you to do.

I may use copies of your work for in-class discussion; when I do so, I will not identify the author publicly.

The grade you earn on your journal will count for **one-third** of your semester grade.

REFERENCES

1. Mathematical Sciences Education Board et al., *Everybody Counts: A Report to the Nation on the Future of Mathematics Education*, National Academy Press, Washington, DC, 1989.

2. M. Van Doren, *Liberal Education*, Beacon Press, Boston, Massachusetts, 1959.

3. J. Mason et al., *Thinking Mathematically*, Addison-Wesley, Wokingham, England, 1985.

COURSE SPECIFIC

but with

BROADLY APPLICABLE IDEAS

Writing Assignments and Course Content

Joanne Erdman Snow
Saint Mary's College, Notre Dame, Indiana

The mathematical community has recognized that there is a need to integrate writing assignments into our courses. This awareness presents two new issues: overcoming student resistance and finding natural assignments which enhance the learning of the subject.

Overcoming student resistance is generally the easier of the two problems. Moreover, the resistance usually comes more from students who are majoring in mathematics than others. As one student wrote on her evaluation of me at the end of a term, "I became a math major so I wouldn't have to write papers." However, mathematicians do write, so it isn't unreasonable to expect the students in mathematics courses to write. Reminding students that in their profession, whatever it may be, they will have to write various kinds of reports and letters is generally sufficient to convince them there is some long-term benefit to these writing exercises. Moreover, a little experience with the writing assignments often shows them that they are learning the material better. So there is immediate payback for them.

"I became a math major so I wouldn't have to write papers." However, mathematicians do write, so it isn't unreasonable to expect the students in mathematics courses to write.

The problem of finding assignments which are natural, which enhance the course and the learning process, and which do not seem imposed merely to get the students to write is the more difficult of the two. A writing assignment can be designed to satisfy one of three purposes. In the course of explaining them, I will provide some examples from my experience.

Everyday Writing Assignments

The first and most obvious purpose is to reinforce the day-to-day material covered in class. Everyday homework assignments can serve as mini-writing assignments. From the start of the course, it is helpful to require or strongly suggest that students use words when they solve a problem or write a solution. A string of calculations which made sense Monday night can be meaningless a week later. Using words helps students to learn and organize the material, as opposed to just mastering a recipe. One can make short assignments based on a theorem or definition discussed that day. For example, asking the student to explain and provide examples of the various ways the definition of continuity can be violated is an exercise which helps the student not only learn the definition but examine all the implications of the simple statement that "f is continuous at x_0 if $f(x) \to f(x_0)$ as $x \to x_0$." A third idea is to have the students keep a journal for the course (see, for example, [4], [5], [6]). The journal can serve various purposes. Students can include a short summary of the material taught that day, questions for the teacher, personal comments as to how the course is going, and even sample test questions devised by the student. One can sell the journal by indicating that it will be a great study guide for the test. In-class journal writing can also replace quizzes and would probably be a more meaningful experience for the student [5].

Assignments with a Broader Goal

The above assignments are rather local in nature. Assignments can also be used as learning devices in another broader or more global sense, namely, to have the student take a step back and look at the content of several class lectures. These types of assignments help students to digest the course material or make it their own. I will give four examples of these and explain my objectives for each.

The first example is an exercise of the form: write everything you know about a particular subject. My goal in giving such an assignment to the students is to have them pull the bits together, thereby learning the bits as well as how they fit together. For example, the gradient is a topic which we spend several lectures on in Calculus III. It is a rich and powerful notion that one would hope the students would appreciate as more than just a tool used in certain kinds of problems. This idea can be adapted in many classes; other topics are applications of the derivative or integral, the meaning of the determinant, uses of the fundamental theorem of arithmetic, implications of Cauchy's integral theorem, countability, tests for extrema [3], techniques of integration [3], or the various kinds of statistics.

Sometimes, it is necessary to be more specific in making an assignment. Giving the students a list of questions to guide their thoughts about a particular topic is helpful. One can also get the students to stretch their brains by asking questions which go beyond the standard material for the class or homework. The topics of partial derivatives, differentiability, continuity, and the relation between the three lend themselves to such an assignment. Here one could pose such questions as:

> Does the existence of partial derivatives guarantee continuity?
>
> Does the existence of partial derivatives guarantee differentiability?
>
> Does the existence of directional derivatives in all directions guarantee differentiability?
>
> How is the gradient like the derivative of a function of one variable?

Other topics which suit this format are the development of the notion of the derivative or integral. Students may follow the explanation in class, nodding their heads as you present the next step or even making suggestions, but this may be the limit of their understanding. When they enter physics class and it should be obvious that a certain calculation requires the use of a limiting process equivalent to the integral, they may not recognize this. Or in Calculus III when one defines the double and triple integral, the students will probably not remember the ideas of partitions and lower and upper sums. So a set of questions which helps them write out the derivation of these notions is beneficial in helping them understand and retain these ideas.

Another source for exercises which appears in any math class is the infamous "if . . ., then" statement. One can ask if a given "if . . ., then" statement is true and then if the converse is true. To answer these questions students must come up with examples or theorems to back up their claims. This helps them present and understand arguments. These exercises could also be part of an exam. "Thought questions" which require more than a numerical or symbolic answer force the student to write and gather their thoughts. As this may be a new experience for the students it is often helpful to warn them that this type of question will appear. As the move towards a "lean and lively calculus" emphasizes content as opposed to routines in our math classes, these types of questions will be appearing on our tests naturally.

The last example is meant to remind the student of the purpose of a subject. In differential equations, students learn several techniques for solving equations. They can get lost in the recipes and forget what they're cooking. The assignment given the student is to find a real-life problem, model it mathematically, and then solve and interpret the solution. This assignment reinforces the idea that a differential equation is a mathematical statement of a problem and that meaning must be attached to the solution in order for it to be of use. Doing projects in a statistics, math modelling, or graph theory course would serve a similar end.

Assignments to Enrich the Course

Continuing in this mathematical terminology, a third purpose an assignment can serve is to lead the student on a path tangential to the course content. Here, too, the teacher has many options for kinds of assignments. He/she may require the students to develop something which was touched upon in the class or one of those "optional topics" in the text. For example, last year when we talked about series and their role in approximating numbers and functions, the students were asked how they know pi is 3.14 . . . Their assignment was to look up the history of this number. While this material would not be part of a test, what they did find out showed them a real-world application of a topic discussed in class. It was with great pleasure that at the end of the year, I could show them an article in the most recent edition of the *Mathematics Magazine* [2] on this very subject. For teachers of real analysis, Bartle [1] has several projects sprinkled throughout his text. A carefully designed set of questions or statements can guide the students through the development of some nonessential, but interesting, topic which one would like to cover in class, but for which there is not enough time.

Too often our students get discouraged . . . Once they discover that the great mathematicians also struggled, the students may be more encouraged, motivated, and fascinated by the field.

In this same vein, students can write about the history of a particular subject. From this they gain some appreciation for the topic and, more generally, some experience with the idea that mathematics is not a stagnant field. They might find it amazing to know that complex numbers were treated with great distrust for a long time, that the notion of series evolved over many centuries, that the current definition of the derivative was not written down in a matter of minutes on one fine afternoon. There is an added benefit to this historical study. Too often our students get discouraged, thinking that everything should be as easy and obvious to them as it was in high school. Once they discover that the great mathematicians also struggled, the students may be more encouraged, motivated, and fascinated by the field.

Another way to enrich the course or the students' understanding of mathematics is to have them read a book about the nature of the subject. Mathematics has some obvious examples in Hardy's *A Mathematician's Apology* and Halmos' *I Want to Be a Mathematician*. These can be appreciated by all students. Sending the students to the library to read a journal article (for example in *Mathematics Teacher, The College Mathematics Journal,* and *The Mathematics Magazine*) can also help them reach beyond the well-defined line of the course content. In making this an option in liberal arts courses as well as in an upper division course, I have learned many interesting things about map-making, the structure of molecules, mathematics education, various mathematicians, the architecture of domes, astronomy, and even the personality of my

students. From student comments in essays on such articles, it is clear that their knowledge of what mathematics is has expanded. Non-majors have come to appreciate the subject, even if they don't like it.

Tips on Presenting the Assignment to the Students

I would like to pass on a few tips about making assignments. An assignment can never be too well-explained. When preparing it, try to anticipate some of the students' questions about what the teacher really wants them to do. When asking them to summarize a topic covered in class, one might suggest the use of outside sources to avoid receiving a regurgitation of their classnotes. Put on reserve books for a topic for which there might not be many sources. Make the intended audience clear.

An aspect of writing which is not directly related to the content of the assignment but is essential for learning is the process of revision. While it is impractical to have every assignment go through countless revisions under the supervision of the teacher, it is important that some assignments be revised. Unless students are forced to go back and say something correctly, they may never get the idea straight. If the goal of writing in mathematics classes is to help the student learn and we merely say "Your statement that continuity is equivalent to differentiability is incorrect," then we have not really helped the student know and understand the truth. To ease the burden on the instructor in this regard, one can make use of peer evaluation. Requiring that a classmate read the paper before it is resubmitted to the teacher should be beneficial to both students and a time-saver for the instructor.

It has been my experience that both the students and I have found these writing assignments interesting and rewarding. I keep this in mind when I stare at a stack of thirty or forty papers and wonder what possessed me to make an assignment which once sounded so wonderful to me.

REFERENCES

1. Robert Bartle, *The Elements of Real Analysis,* second edition, Wiley, New York, 1976.

2. Dario Castellanos, "The Ubiquitous π," *Mathematics Magazine* 61 (1988): 67–98.

3. Marvin L. Johnson, "Writing in mathematics classes: A valuable tool for learning," *Mathematics Teacher* 76 (1983): 117–119.

4. Coreen L. Mett, "Writing as a Learning Device in Calculus," *Mathematics Teacher* 80 (1987): 534–537.

5. Cynthia Nahrgang and Bruce Petersen, "Using Writing to Learn Mathematics," *Mathematics Teacher* 79 (1986): 461–465.

6. M. Watson, "Writing has a place in a mathematics class," *Mathematics Teacher* 73 (1980): 518–519.

Library and Writing Assignments in an Introductory Calculus Class

John R. Stoughton
Hope College, Holland, Michigan

During the last several years the mathematics community has engaged in a great deal of discussion concerning the teaching of calculus (e.g., the MAA conference reports, *Toward a Lean and Lively Calculus* and *Calculus for a New Century*). Over the last twenty or thirty years the emphasis of the calculus course has shifted from understanding concepts to problem solving. Listen, for example, to Lynn Steen, former president of the MAA:

> [Freshmen courses should] help students learn to think clearly, to communicate, to wrestle with complex problems.
>
> There is very, very little in the calculus course of today that does any of these things. Word problems are a small step in that direction, but they are rare ...

Thus while we once taught concepts in the classroom and hoped that, as a by-product, students would learn to solve problems, today we teach problem solving and hope that eventually students will learn the theory behind it all. Unfortunately it rarely works. Now the question is "How do we reverse this trend?" I believe that an answer may come in borrowing an idea from our colleagues in the humanities—if you want students to understand something, have them write about it.

If you want students to understand something, have them write about it.

During the fall of 1988 I taught two sections of introductory calculus in which students were given library and writing assignments. Each student was given two assignments, the first concerning a topic from precalculus and the second on a topic in calculus. They were instructed to research their topics (via either textbooks which I had put on reserve in the library or their own high school or other texts) and then to paraphrase their research in a term paper which was to be written in a clear, concise, readable fashion. I emphasized that it must be technically correct. They were also to write a paragraph critiquing each of the textbooks from which they read.

There are several reasons why I chose their first assignment to be on a topic from precalculus. First, I wanted them to do it early in the semester, which meant that it would be before most of them had been introduced to such calculus topics as limits or derivatives. Second, I wanted them to write their first paper on a topic they felt they already understood, and hence a topic with which they did not feel too uncomfortable.

During the first day of classes I told the students about the two assignments they would be given and briefly described what would be expected of them. At the end of the first week I handed out the following list of topics from which I (not they) would choose their assignments:

1. Completing the Square and the Quadratic formula. Include a short discussion of the method of completing the square, and then show how that method is used to derive the quadratic formula as the solution(s) to the general quadratic equation, $ax^2+bx+c = 0$. Also include a discussion of the discriminant and its relation to the roots of a quadratic equation.

2. Inequalities. Discuss the properties of inequalities and the interval notation for solutions to inequalities. Outline a general method for solving quadratic (second degree) inequalities. Be sure to include a discussion of the difficulties involved in multiplying or dividing an inequality by an unknown quantity or expression.

3. Functions. Give the definitions of a relation and a function (give examples to illustrate the difference). Define and discuss the domain and range of a function. Discuss composite and inverse functions.

4. Systems of Equations. Define a linear equation in two unknowns. State and illustrate a general method for solving two linear equations in two unknowns. Discuss why this method might not work for nonlinear equations.

5. Absolute Value. Give the definition of absolute value. Discuss some of the properties of absolute value (e.g., $|ab| = |a| \, |b|$). Also discuss the relationship between absolute value and radicals. Tell how absolute value is related to distance on a number line.

6. Symmetry and Graphing. Give the tests for symmetry with respect to the x-axis, the y-axis, the origin, and the line $y = x$. Discuss the geometric significance of each of these symmetries. Give examples of each.

7. Transformations, Translations, and Stretchings. Discuss vertical and horizontal translations; vertical and horizontal stretchings. Compare the graph of $y = f(x)$ to the graphs of $y = af(x)$, $y = f(ax)$, $y = f(x - b)$ and $y - b = f(x)$ for real numbers a and b.

8. Equations of Lines. Discuss the equations of horizontal and vertical lines, the point-slope form of the equation of a line, and the slope-intercept form. Also give a short discussion of parallel and perpendicular lines.

9. Factoring Trinomials. Discuss your favorite method of factoring a trinomial into the product of two binomials. Also, state the factor theorem and tell how it is related to the problem of factoring a polynomial expression.

10. Slope of a Line. Give the definition of the slope of a line and discuss why it doesn't depend on which two points you choose. Also give a discussion of the interpretation of a large, positive slope; a small, positive slope; a small, negative slope; and a large, negative slope.

11. Rational and Irrational Numbers. Define rational and irrational numbers. Prove that $\sqrt{2}$ is irrational. Which rational numbers have decimal expressions that terminate? Which ones repeat?

12. Synthetic Division. Discuss and illustrate this method of division of polynomials.

A week or so later I made the individual assignments and encouraged them to begin work. The topic assignments were made somewhat at random, although by that time I was beginning to recognize which students were roommates or close friends and I tried to avoid giving them the same topics. Ideally, I would like to have been able to assign a different topic to each student, but obviously there were far more students than topics.

The students were told to write an article of from two to four typewritten pages on their assigned topic. I emphasized that the article should be written in the student's own words, paraphrasing but not copying the texts that they read. I encouraged them to try to re-member the problems or frustrations they may have encountered when they were first introduced to the topic in high school, and to write as though they were writing a portion of a textbook with the aim that other students be able to read and understand it. I encouraged them to think of a younger brother or sister who might be studying the topic this year, and to ask themselves what they might write to help that younger sibling gain a deeper insight and understanding of that topic. I also encouraged them to supplement their articles with examples, graphs, or illustrations as long as they were sure to label carefully. Finally I reminded them to write a para-graph reviewing the section of each textbook from which they read. I emphasized that they should discuss what they liked as well as what they disliked about each book.

> **I encouraged them to think of a younger brother or sister who might be studying the topic this year, and to ask themselves what they might write to help that younger sibling gain a deeper insight and understanding of that topic.**

I must admit that at first I was apprehensive about grading their pa-pers. I have taught introductory calculus enough now that I feel very comfortable about making examinations for any and all parts of the course and about assigning grades to the results. However, grading term papers was a whole new experience—one for which I did not feel particularly well prepared. But while I did find myself agonizing over some of the papers (how much original thought was involved here? how much credit should I give there?), by and large the grad-ing went smoothly. While most of the students took the assignment seriously and did a good job, some of the papers obviously had been done the night (perhaps even the hour) before class. For example, four students from the two classes were assigned the topic of factoring trinomials. Three of the four wrote a page or so carefully outlining the "FOIL" method that they had learned in high school, supporting it with two or three examples, then outlined other methods they read about in precalculus books, discussed the more general problem of factoring polynomials, stated the factor theorem, and discussed, via examples, how it is used in factoring. They con-cluded with one or two paragraphs of criticism of the books they had read. The fourth paper, however, which was about one and a half pages in length, quickly discussed the "FOIL" method (no examples) with several incomplete sentences and even more in-complete thoughts. There was no discussion of the factor theorem and no indication of which, if any, books he had read in researching the assignment. It was not difficult to decide the relative merit of that paper.

Another problem in grading these papers was that the assignments themselves varied in difficulty. For example, the students who were assigned to write about rational and irrational numbers encountered more problems than the others. With a little help they were all able to read and do a reasonable job paraphrasing a proof that $\sqrt{2}$ is irrational, but none had a guess as to which rational numbers have decimal expansions that terminate and which ones have expan-sions that repeat. At the other end of the spectrum, I only included the topic of synthetic division because I had run out of ideas. I knew in advance that it would be difficult for students to express any cleverness or originality with this topic, and I had to keep that in mind as I graded those papers. I should say, however, that one of the best papers I received was written on that topic. All in all the reading and grading of these papers did take up a fair amount of extra time on my part, but the grading was not as difficult as I had anticipated.

The second assignment was made much later in the course. After about the tenth week of the semester I handed out the following list of topics:

1. Limits. Give the definitions of the limit of a function, one-sided limits, and infinite limits. Give your own pedagogical description of what a limit is and how to find it. Use the defini-tion to give a careful proof of a limit problem (e.g., prove that $lim_{x \to 2}(3x + 1) = 7$).

2. Continuity. Give the definitions of continuity of a function at a point and on a set. Give your own pedagogical description of the meaning of continuity. Use the definition to prove that the composition of two continuous functions is continuous.

3. Tangent Lines. Give your own pedagogical descriptions of secant lines and tangent lines. Give the definitions of the slope of a secant line and the slope of a tangent line. Show how this leads naturally to the study of the derivative.

4. Instantaneous Velocity. Give your own pedagogical descrip-tions of average and instantaneous velocity. Then give the definitions of each and show how this leads naturally to the study of the derivative.

5. Derivatives. Give the definition of the derivative. Give several examples. Include a discussion of circumstances under which the derivative fails to exist. State and prove the product rule for derivatives.

6. Relative Extrema. Give the definitions of (relative) maximum and minimum values of a function. Discuss increasing and decreasing functions. Then give a general discussion of how to find maximum and minimum values for a function.

7. Concavity. Define the concepts of a function being concave upward and concave downward. Give a general discussion of how to find where a function is concave upward, concave downward and points of inflection.

8. Chain Rule. Review the definition of the derivative and then state and prove the chain rule. Also give several examples of how it is used.

9. Differentials. Give the definitions of dx and dy. Discuss the difference between Δy and dy. Give examples to show how dy is used to approximate Δy.

10. Mathematical Induction. State the principle of mathematical induction and give your own pedagogical interpretation of its meaning and why it is true. Use it to prove that $1^2 + 2^2 + \cdots + n^2 = \frac{n^3}{3} + \frac{n^2}{2} + \frac{n}{6}$.

11. The Mean Value Theorem. State and prove Rolle's theorem. State the mean value theorem and give your own pedagogical description of its meaning. Illustrate with examples.

12. Least Upper Bound Axiom. Give the definitions of upper and lower bound for a set of numbers. Give examples of sets which are and are not bounded above. Define least upper bound and greatest lower bound for a set of numbers. Give examples of sets which do (do not) contain their least upper bound. Also state and discuss the Archimedean property of real numbers.

A week or so later I made the individual topic assignments and they were given two weeks to complete their papers. Since by this time we had covered each of these topics in class and they had been tested over that material, I attempted to make topic assignments coincide with weaknesses as observed on their tests; except that again I tried to ensure that roommates, friends, or others who might be tempted to work together were assigned different topics.

In my oral instructions to the students I emphasized that while these (second) papers need not be any longer than the first, I did expect them to research their topics much more thoroughly. I also stressed that although I again wanted them to paraphrase and not copy from the textbooks they read, it was very important that their writing be technically correct. I warned them that this would be graded much more strictly than in the first paper.

Indeed, I did observe significant improvement in the quality of writing in the second assignment. Not only quality, but quantity as well. Most of the papers were quite a bit longer than I had required. The average length was more than six pages. But what I found most interesting was their written evaluations of the textbooks from which they had read. In the first paper the vast majority of the criticisms were positive—some even highly positive. However, in the second assignment many were very harsh in their criticisms of the first book from which they read and progressively less so with the second and third texts. In fact, many had great praise for the last text. Since I had put only eight calculus books on reserve at the library, and since each student selected them more or less at random (and in random order), the same book was severely reproved by some and highly praised by others. In retrospect, I believe that most students had difficulty understanding what they were reading the first time they read it, and held the author of their first text accountable. Then as their comprehension began to increase, the second and third texts miraculously became more readable.

At the end of the semester I asked each student in these introductory courses to write a paragraph giving their reaction to the writing assignments. In general their comments were very positive—some even admitted to having learned something. For example:

  I liked the idea, although I'd never heard of writing a paper in math before. I learned a lot from it. I think it's a good idea.

  . . . it helped us understand the material better—and it wasn't as if they were big, huge term papers as in English and such.

  At first I thought a term paper for calculus was a stupid idea. It's still not my favorite part of the class, but it did help a bit . . .

  When I first heard about term papers in a math course, I felt that it was first impossible and second pointless. However, the papers I wrote have given me a much better understanding of the concepts we studied. Up until this course, I did my math by memorizing formulas and equations without really knowing what was actually going on. Now that I've had to explain a formula . . . I've really learned the reasons behind the equations. In fact, I've told my high school teacher about this idea and he's seriously considering it.

I must confess, however that not all the comments were positive. Witness the following:

  I think that the papers were unnecessary and that the students' time should be used to solve calculus problems instead. I did learn from the papers I wrote, but I don't think it made much difference in learning to solve calculus problems.

This project was not intended to have measurable short-range success. In fact, the final grade point averages from these two classes are not significantly different from other introductory calculus classes I have taught. This is not surprising to me. The purpose of the assignments was to encourage students to think more deeply about certain concepts; the written tests still emphasized problem solving. The real test of the effectiveness of this project will come when and if these students decide to become mathematics majors and take our upper-division courses. Hopefully, their transition from the calculus and other lower-division courses will be much smoother than is the experience of most of our students. However, in order that this smooth transition become a reality, I believe it is important that we begin teaching students to write mathematics at all levels of the curriculum for the major. These types of writing assignments can be included in all levels of calculus courses, not just the introductory course. Also, they could be implemented in such sophomore courses as linear algebra, discrete mathematics, and differential equations.

Teaching Mathematics Within the Writing Curriculum

David T. Burkam
University of Michigan, Ann Arbor, Michigan

> . . . one can be a strict logician or grammarian, and at the same time full of imagination and music.
> Hermann Hesse, *The Glass Bead Game*

> Very simply, poetry and mathematics are two very successful attempts to deal with ideas.
> Scott Buchanan, *Poetry and Mathematics*

Introduction

Over two years have elapsed since I first began to actively create a course on the lesser seen faces of mathematics. Five years of teaching college calculus had been satisfying, but limited in terms of success. Yes, most of my students learned the mechanics of the techniques, some of my students learned the appropriateness of and motivation behind the techniques, but hardly anyone felt the passion and beauty of the techniques. My own enthusiasm was as much suggestive to them of personality defects within their instructor as of some previously unexperienced, artistic realm lurking behind the tedium of formulas, theorems, and story problems. Then an opportunity arose whereby I might have a reasonable chance of presenting a defense of mathematics—and I eagerly accepted the challenge. This short article is the story of that attempt, written in the hope that others might try similiar or even entirely different approaches until we can eliminate the misconceptions about our discipline.

My seminar was part of a larger program whose goals need to be known in order to understand my small contribution these last two years. Within the large University of Michigan structure, a small gathering of students and faculty have spent over 20 years creating a degree-granting, liberal arts subdivision known as the Residential College (R.C.). Nearly 800 students and over 50 professors and lecturers are currently keeping the dream alive—a dream of small classes, high interaction, personal attention, and interdisciplinary academic excellence. The R.C. is not an honors program, but a college enjoying all the benefits of a large research institution while maintaining the student orientation and broad academic emphasis of a small liberal arts school. Central to the philosophy of the R.C. is the First Year Seminar Program, our own version of freshmen composition. Limited to 15 students per section, these seminars provide students with an introduction to the collegiate setting, while emphasizing the analytic skills of writing necessary for undergraduate success. Effective writing, however, requires a substantial subject, and critical writing requires exposure to other people's ideas. Hence, each seminar focuses on a theme, an issue, most often in a manner unavailable elsewhere in the University.

I had spent four years on loan from the Department of Mathematics teaching standard sections of calculus within the R.C., and had come to love the students, faculty, and endless possibilities of the college, yearning to involve myself in the true spirit of the setting. During the fall of 1987, I finally did so by offering my own First Year Seminar: "Euclid Alone Has Looked On Beauty Bare"—Topics in Mathematical Thought. I borrowed my course title from Edna St. Vincent Millay's sonnet, having run across it years earlier and woefully wondered how much sense it would make to the average college student.

The Seminar

The joy of creating your own course includes the opportunity to compile a favorite reading list, ordering ideas in your own way, emphasizing whatever is of most interest to you at that time. Eventually, nearly 20 different books contributed various pieces to the seminar, providing a source for the breadth of ideas crucial to the ongoing discussions and writing assignments. These readings fell into a loose progression of three stages: 1) a general introduction to and exploration of mathematics, 2) a sketch of the historical developments in geometry through the non-Euclidean discoveries into today's notions of time, space, and dimension, 3) an introduction to axiomatic systems, Gödel's results, and the philosophy of mathematics.

The first third of the couse was the eye-opening one, where we dared to suggest that mathematics, like other disciplines, involved subjective elements and inspired debates. It was something more than an endless succession of numbers and equations to be memorized. Sections of the Davis and Hersh book, *The Mathematical Experience*, as well as Newman's *The World of Mathematics*, provided plenty of fuel for student discussions and writings. After an initial journey through texts by Plato, Paul Halmos, Norbert Weiner, and others, this section culminated with G. H. Hardy's *A Mathematician's Apology*, incorporating all of the ideas of aesthetics, motivation and the "pure-versus-applied" debate. In those few short weeks, I had my students hooked. Even the confirmed math haters were finding themselves defending the subject to roommates, relatives, and friends. They were only beginning to get a glimpse of that different world, yet enough of a glance to want more.

We dared to suggest that mathematics, like other disciplines, involved subjective elements and inspired debates.

The middle portion intentionally focused upon a branch of mathematics whose foundation is so common to the high school curriculum that I was sure no college student would be without it—geometry. (Although a few of my students had little beyond their geometry and algebra II coursework.) The often counter-intuitive axioms and theorems of Lobachevskian and Riemannian geometry preceeded further destruction of spatial prejudices as we unlocked our vision through Abbott's *Flatland*, Burger's *Sphereland*, and Rucker's *The Fourth Dimension*. Mathematics had become a source of wonder, the springboard for the imagination and not its bitter enemy.

Finally, building upon a more mature understanding of geometry, axiomatic systems returned our discussions to the formal nature of mathematics, only to disrupt the effort by Gödel's incompleteness results. By now, most of the students had independently adopted one of the traditional philosophical outlooks without knowing yet the specific labels, and the fledgling Platonists, Formalists, and Constructivists argued as fiercely as their professional counterparts!

The Writing Assignments

Despite the extensive reading, the seminar was primarily conceived as a composition course, an opportunity to impart the necessary writing skills for future college exposition. I freely admit that this was a role which brought me no small portion of discomfort and self-doubt. My training had never been in composition, and except for peer tutoring as an undergradute, all of my pedagogic experience centered around mathematics. Who was I to teach freshmen writing? Then slowly I realized that just such an attitude reflects so well our current segregation of disciplines and skills, where as a mathematician or scientist one need not bother to acquire the ability to write, while as a humanist one need not be concerned with the technical knowledge of our time. If I were to be successful in changing my students, I first had to change myself. This unshackling of ourselves from the superficial and harmful restraints of these intellectual boundaries seems to me to be the single most important challenge facing all of us involved with education at any level, especially beyond the high school years. Students need to see in us, and develop within themselves, the ability to connect all intellectual activities, without regard for the accidental divisions that have emerged. I knew I could not replace formal expertise and training with simple dedication, but I also realized that this seminar would be only one of many opportunities for these students to work on their writing. I didn't have to do it all. I wasn't alone.

Who was I to teach freshmen writing? Then slowly I realized that just such an attitude reflects so well our current segregation of disciplines and skills, where as a mathematician or scientist one need not bother to acquire the ability to write, while as a humanist one need not be concerned with the technical knowledge of our time.

The actual writing assignments were of three types: regular short essays, journal entries, and one major research paper. The variety kept all of us working without creating too much of a routine. One 3 to 5 page paper was due approximately every other week, while journals were submitted on the alternate weeks. The final weeks of the semester provided the time for the outside reading and writing necessary for the longer term paper. We never used a writing text, but during the seminar's second run, I incorporated Richard Mitchell's *Less Than Words Can Say* into the reading list as a way to force discussions about language and precision. Professor Mitchell's controversial book changed the tenor of the experience tremendously. His concern over the deterioration of language and education caused an increased awareness in my students of their own lack of clarity and organization, and assisted in the destruction of the supposed differences between mathematics and language. This goal became uppermost in my mind during those months, with more success than I thought possible.

Surrounded by a wealth of ideas, most of which were quite new and startling to my students, essay topics were plentiful. For many the difficulty lay in selecting from the wide assortment. Only two topics were assigned. Later, I made suggestions, but always encouraged creativity and required students to decide upon their own subject. Before any reading or discussion, each student wrote in response to "What is mathematics?" No guidelines. No further specifics. And the results were fascinating, and depressing for what they suggested about our schools. Two weeks and several readings later, the second essay was due. Following in the spirit of Plato's "being" and "nonbeing" distinction, where one cannot know what a thing is without knowing what it is not, the students discussed "What mathematics is not." A much more difficult task, to be sure, especially

as the answer to the first question was no longer clear to them. Subsequent papers were left to the students' interests and the material covered so far. This freedom required more effort on their part, but ultimately produced better writing and heightened organizational skills than a more strictly regimented approach might have done. Does G. H. Hardy embody the characteristics of Davis and Hersh's "Ideal Mathematician?" How can one reconcile mathematics as an abstract activity with its constantly surprising application in the physical world? Why, for centuries, were mathematicians unable to accept the growing evidence that Euclid's parallel postulate was not provable from his other assumptions? Issues of scientific creativity, elitism, value, socialization, and reality intertwined in their writing as both their understanding and appreciation of mathematics improved along with the ability to express these new found ideas in their prose.

The one longer term paper, due at the very end of the fifteen week semester, required additional outside reading, and went through the usual stages of outlines and drafts. We read so much, so quickly, that in-depth understanding was spotty. Here was the opportunity to gain knowledge of each student's choice. Interest and mathematical background dictated much of the selection process, but an impressive array of topics always emerged: fractals, Kant's philosophy of mathematical truth, the search for a unified field theory, artificial intelligence, parallels between mathematical and artistic movements, ethics and morality in research, the interplay between culture and mathematics, and others. Everyone learned from these papers, myself included, and their collective quality was a testament to the untapped potential in so many young people. While only a small handful of the thirty students from the two years have gone on to science or mathematics related coursework, they left with new knowledge and respect for a discipline that woefully so often imparts little of either in introductory courses.

I intentionally reserved mention of the journals until the end, for I wish to stress their unique contribution to the success of the seminar. Formal writing remains a difficult task for most of us (and I include myself in that group). One must be clear and concise, with prescribed goals to be worked toward and achieved. Journal writing, however, encourages early expression of ideas, before they are crystalized into orderly gems of wisdom. One can be fanciful and adventuresome in such a setting, exploring new possibilities without justification or transition. As much of our reading was difficult at times, I was pleased to see the students wrestling with the ideas on paper, asking questions, and providing self-feedback on their progress. While face to face discussion with an instructor, especially for younger college students, inspires certain fears and discomforts, regular communication through informal writing eased those tensions. Possible criticism and evaluation were never an issue here. Effort and honesty were the only requirements. I learned more about my students and their individual needs from the journals than from any other interaction.

While face to face discussion with an instructor, especially for younger college students, inspires certain fears and discomforts, regular communication through informal writing eased those tensions.

So often a final written piece fails to capture the quality of the preceeding effort. We tend to teach students, even by our own example, to hide away the paths and reveal only the destination. Through journals, these dangers can be avoided. One student, whose journal entries were consistently innovative and probing, wrote formal essays that rarely rose above dry, commonsensical notions weakly presented. As the weeks passed, I saw little improvement with his essays despite the interesting ideas in his journal. When I finally confronted him with this disparity, he disclosed a long history of writing difficulties and the strictly-enforced rigidity of prose format in his eduation. He felt the need to write in, what was for him, an unnatural style with words not of his own choice. Not so with the journal where he enjoyed the freedom from convention. I encouraged him to approach his next short essay as if it were simply a polished journal entry, in his own language. The result was a much stronger essay, with a clear voice and direction. By the end of the term, he tackled a difficult collection of readings, sophisticated both in style and content, and wrote an excellent paper. Had it not been for his journal writing, I would never have guessed his true ability.

Conclusion

Success is such a relative notion. Most of my students felt their writing improved, and most displayed a new respect for mathematics and language. Since there were no formal prerequisites for the seminar, the math background spanned a broad continuum of exposure, from little more than geometry and advanced algebra through advanced calculus. Assuredly, the individual understanding of some topics was a function of this background, but I found such diversity appealing (and challenging). Rather than restricting, the heterogeneity of the students enhanced the quality and scope of the seminar. The students themselves capitalized on each other's knowledge and were thrilled by a discussion format where Ayn Rand and the theory of aesthetics were as likely to be mentioned as Stephen Hawking and the mysteries of black holes. I certainly owe much of the success of the seminars to the students and their willingness to attempt almost everything.

Traditional mathematics appreciation courses often result in neither of the two, mathematics or appreciation. Some history and philosophy of mathematics courses occur so late into the curriculum that very few students are made privy to such ideas. I am searching for that elusive compromise where the access is as open as the desired ends. Where arbitrary boundaries of thought can be separated into the meaningful and the meaningless distinctions. Where the mathematical language and the English language can be seen in their similarities as well as their differences. I can make no claims of achievement except upon a small and preliminary scale. But I hope others attempt likewise. We are blessed with an ever-growing number of writers among us whose published books can provide educators with the necessary tools to accomplish these changes. My efforts would have been futile without their hard work. I intentionally avoided a detailed account of my reading list as selections were based upon my own prejudices and interests, but I urge the community to make better use of the increasing number of literary treasures. Their very existence attests to the successful marriage of mathematics and expository writing.

In closing, I defer to Edna St. Vincent Millay. No one is surprised when a mathematician values her own craft, but when an outsider so eloquently expresses such reverence, there must be something there:

Euclid alone has looked on Beauty Bare.
Let all who prate of Beauty hold their peace,
And lay them prone upon the earth and cease
To ponder on themselves, the while they stare
At nothing, intricately drawn nowhere
In shapes of shifting lineage; let geese
Gabble and hiss, but heroes seek release
From dusty bondage into luminous air.

O blinding hour, O holy, terrible day,
When first the shaft into his vision shone
Of light anatomized! Euclid alone
Has looked on Beauty Bare. Fortunately they
Who, though once only and then but far away,
Have heard her massive sandal set on stone.

Edna St. Vincent Millay

Writing About Proof

Keith Hirst
Southampton University, Southampton, England

Introduction

The notion of proof is, of course, central to mathematics. In lectures and problem classes we spend a lot of time demonstrating proofs to our students. But what do the students think a proof actually is; what is the nature of the procedures; how do you know when you have a proof; how can you tell whether it is correct, or complete? These are just a few of the many questions about the nature of proof which are usually not addressed at all in most courses. As a major aspect of mathematical activity which distinguishes mathematics from other areas of scientific enquiry, it is surely important that students should have some understanding of the nature of proof. Perhaps it is felt that this will come about by the fact of their being exposed to so many proofs throughout their mathematical education. How can we test this assumption? An obvious method is to ask the students to write an assignment specifically concerned with proof. Having seen many of these over the years, it is my experience that an invitation to write an essay on "Proof," whether arising from a lecture course on the nature of mathematics or not, invariably gives rise to a succession of generalities, concerned with such notions as truth, certainty, rigour, logic, etc., but with no exploration of the students' own personal perception of proof. What I want to do in this article is to describe a rather different approach, based on a proof constructed by some students in the context of a particular problem. This gave rise to a discussion about the particular proof offered, which incidentally shed light on the students' general attitudes to proof.

The Problem

bases	ten	nine	eight	seven	six	five	four	three	two
squares	1	1	1	1	1	1	1	1	1
	4	4	4	4	4	4	10	11	100
	9	10	11	12	13	14	21	100	1001
	16	17	20	22	24	31	100	121	10000
	25	27	31	34	41	100	121	221	11001
	36	40	44	51	100	121	210	1100	100100

Figure 1

The situation which gave rise to this work took place in the context of an investigation-based course (see [1] for more details), where most of the students were intending to become high school teachers, but where the course was a component of their mathematical studies. They were presented with the above partial table of square numbers in various bases and asked to investigate the patterns they observed, to describe them and to prove their conjectures. Many patterns emerged, but the one that caught the attention of the class involved the diagonals, and the form of the number $(n-a)^2$ in base n. A group of three students was asked to present their solution to the rest of the class.

The Students' Proof

Lynn presented the group's result and proof as follows. She copied the table (Figure 1) onto the blackboard and labelled some of the diagonals from bottom left to top right. (Diagonal 1 contains all the 100's, diagonal 2 contains 51, 41, 31, ..., diagonal 3 contains 44, 34, 24,..., and diagonal 4 consists of 40, 31, 22, 13,..., so they are numbered from right to left.) She then proceeded as follows.

If we allow the base to be represented by the letter n, then the diagonal labelled 1 can be represented by n^2. In the same way the diagonal labelled 2 can be represented by $(n-1)^2$, the diagonal labelled 3 by $(n-2)^2$ and the diagonal labelled 4 by $(n-3)^2$. The first column of each number in the table, i.e., that nearest to the right-hand side, is a multiple of n^0. The number in the second column is a multiple of n, the third a multiple of n^2 and so on. In this way the numbers can be tabulated:

diagonal	diagonal expansion	table		expansion	limits
		n	$n^0=1$		
$(n-1)^2$	n^2-2n+1	$n-2$	1	$n(n-2)+1$	$n>1$
$(n-2)^2$	n^2-4n+4	$n-4$	4	$n(n-4)+4$	$n>4$
$(n-3)^2$	n^2-6n+9	$n-6$	9	$n(n-6)+9$	$n>9$
From this a general result can be formed for $(n-a)^2$					
$(n-a)^2$	$n^2-2na+a^2$	$n-2a$	a^2	$n(n-2a)+a^2$	$n>a^2$ (*)

Figure 2

So in base n the number $(n-a)^2$ has $(n-2a)$ as its n's digit and a^2 as its units digit (provided $n>a^2$).

I then asked the class as a whole whether they thought this was a *proof* of the result. This gave rise to a fairly heated discussion, for which I was quite unprepared, having on other occasions tried to start a discussion with the abstract question "What do you think proof is?" At the end of the session I asked the students to go away and write an assignment about proof in the light of the discussion they had just had. An account of these assignments follows, largely composed of extracts from the students' own writing—it speaks for itself. One interesting observation is that at the end of the class discussion, during which opinions as to whether or not the proof above was legitimate had been firmly stated, for and against, by class members, they nevertheless asked me to adjudicate. Their question to me was not "do you think it is a proof?" but simply "is it a proof?" Naturally I declined their request!

The Students' Writings

IT IS NOT A PROOF

JILL   I personally did not believe that this one line (*) constituted a proof. I agreed that the results could be generalised in such a way and this generalisation is algebraically correct but if it is considered in the context of the problem it does not adequately prove anything. I believe the algebraic statement provides us with a method of evaluating a specific entry in the table of squares of bases and certainly for all the diagonals we considered it worked perfectly well but a proof must prove some idea or theory totally and I strongly believe this is only a generalisation. It was suggested that the method of proof by induction could be used to convince the "stubborn" members of the class but I don't think the problem could be tackled this way as we cannot discriminate between a and $a + 1$. Any value could be given to either but no satisfactory proof would be achieved. I do not dispute any of the generalised statements put forward by other members of the class. However, I am not satisfied that we have proved our findings.

This is the most articulate piece of writing produced from among the class. Her beliefs about the situation are unequivocally clear. It raises so many issues for us to consider. What kind of impression have we put over concerning proof by induction? What kind of understanding of the use of variables have we managed to impart? The question of a 'specific entry' in the table raises a number of interesting points.

PROOFS ARE USUALLY DIFFICULT

JENNIFER   I felt that this (the last line (*) in Figure 2) was a proof . . . Some students felt that the above was not conclusive proof which led to the question—What is a proof? Where do we draw the line between having a pretty strong theory and having a definite solid proof. And since we are more or less all at the same level of mathematical knowledge—why do we disagree as to whether or not the above is proven? Some of the "unbelievers" muttered the words "induction" and "what about $(a + 1)$?" If the general statement is written out for $(a + 1)$ . . . if we replace $(a + 1)$ by a we get the original equation once again. I think in this case people were held back from believing the proof because the actual "proof" was derived so easily. It was interesting to observe the way in which several of the "proof believing" members of the group tried to convince the "unfaithful" that this was a proof. Intimidation of voice, tone, and choice of words was a major weapon used. Classic opening lines to arguments were words such as "Surely. . ." or "It is obvious that. . ." or "Clearly. . ."I also think that doing long winded inductive proofs and complicated maths has conditioned us into believing that all proofs have to be intricate and extremely clever.

The idea of "proof by intimidation" as a way of describing the use of authority was thought up by one of the students during the discussion prior to the writing. The description stuck and appears in a number of the essays. The interesting thing about Jennifer's account is the observation of the social interactions accompanying proof.

SARAH   Clearly, the expansion of the bracket $(n - a)^2 = n^2 - 2na + a^2$ is algebraically correct, but does this line constitute a proof? Against: (*) is simply the generalisation of observations—it does not show that there do not exist specific cases where this does not hold—therefore it does not constitute a proof. For: In (*), there is no obvious value of n or a where this would not hold as this is clearly an accurate expansion of a bracket—proof needs no more than this.

How is it that in a group of third year mathematicians our understanding of proof is so vague that we can hold such differing opinions about what is meant by such a basic principle.

SARAH was unwilling to commit herself either in discussion or in writing. She shared the views of others about reasons for the confusion however.

Could it be that the unease is caused simply by the fact that from previous experience, proofs are long and arduous, so therefore one line cannot constitute an entire proof. How is it that in a group of third year mathematicians our understanding of proof is so vague that we can hold such differing opinions about what is meant by such a basic principle.

LOGIC AND TRUTH

RICHARD   (Richard embarked on an attempt in his writing to develop a proof in a symbolic logical style. I gained the strong impression that for him the forbidding image of the collection of formal looking symbols was part of what constituted proof. After his attempt, he then wrote as follows)

Of course, you don't want to write down all this logic every time you want to prove something. I think that writing down the intermediate statements and the final statements with a brief explanation of how they are implied from previous statements/assumptions (if not obvious) is all that is required to "describe" a proof. The proof itself includes all the logic as well, but we can do this in our heads, without confusing ourselves with all the details of what the logical steps really are.

SUNITA   If something has been proved it has been shown that it is true. Mathematical truth? Proof is when symbols are manipulated according to mathematical rules to demonstrate mathematical ideas. More than this. (*) is a statement, which is true. On its own it is not a proof, just a manipulation of symbols according to rules of expanding brackets. It becomes a proof when put in the context of bases and when the last step is interpreted as $(n - 2a)$ in the n column and a^2 in the units column.

Unlike the certainty of Jill, this contribution shows clearly the tentative groping towards a means of expressing her ideas. The points about interpretation and context are the important ones.

DEFINITION OR INTERPRETATION?

TAMERA   Like several students, she quoted a 'dictionary' definition of proof, but then continued:

However, many people do not have a dictionary definition of proof in their heads so have obtained their viewing of the word through years of studying mathematics. We have set up particular ways of proving certain types of problems, e.g., induction for natural number problems, proof by contradiction in analysis (e.g., Archimedean Property). So when we can't find a certain way to prove a problem it could be that it cannot be proved, e.g., Goldbach's conjecture concerning primes, or that it does not need one of these proofs because the sequence of events leading to the final conclusion is self-explanatory.

To begin with, like others, she is hinting at the cultural dimension of what is accepted as proof. Her final remarks are interesting in that she appears not to regard straightforward deduction as constituting proof. As with the other writings, the 'ritual' nature of proof ([3] p. 151) seems clearly to emerge.

EVA   People felt that assumptions had been made which made them feel formal proof was needed. I disagreed with that because to me each step of the process was logical and left no room for the need of a proof. The proof is in the logic of each step. The problem I felt that other people had in accepting it as a proof was in the explanation. It came across as if they were saying that because it worked for $a = 1, 2$ it would work for all $a > 0$. This isn't what was really being said. When I looked at this problem I picked out the patterns and then extended the diagonals downwards and checked if the pattern was being followed by converting numbers back into base 10. The idea was given to me that there was a way of explaining why these patterns should exist and this I feel proves that they do exist and is a full and complete proof of why the patterns do occur. It was odd that people should disagree over whether a proof had been given. As we all had similar mathematical backgrounds why should we still not agree over what seems a very fundamental mathematical concept, a proof. Maybe one reason is that we have got used to being given theorems and then a proof by contradiction, etc. In this situation it was just a logical train of thought without any aspect to it that would make it look like the proofs we have got used to being given. In analysis we were shown that even what seems obvious should be proved as it's not always safe to take things for granted. Maybe this is true for some things but not I feel in this situation. Another aspect to this situation is that sometimes we may have come across proofs by intimidation. In lectures one is confronted by an authoritative figure who you feel you must believe and could convince most people that he is "right." Maybe if people were allowed to question proofs more this sort of situation where people are not convinced would occur more often.

One interesting thing about this is that she evaluated the proof offered by working again through the problem. She felt the need for a personal interpretation before making a judgement.

SIMON   This brings us to the problem of how we decide what degree of proof is necessary. In many courses at university the level of proof is determined by the lecturer. This then becomes the level accepted by the student for this course, and these standards will vary from course to course. This means students will accept a proof, as it is given by someone in 'authority' and not tend to question its validity. Proof in its most rigorous form is a definite art within mathematics, whereas the proofs given by lecturers are completely open to interpretation and discussion as to whether or not they are valid, and so the proof of a theorem becomes indefinite.

The interesting thing here is his perception of the idea of different degrees or levels of proof.

JULIA   Therefore the argument is about individuals' interpretation of what proof actually is. Everyone agreed that this general formula is true. No one could find an exception to it but people were still quite reluctant to accept it while others easily assumed that it has been proved. The above example uses very simple maths and algebraic extensions to describe quite a complex relationship between number bases. It is possible that people are looking too hard at the problem and trying to make it harder than it is . . . I don't believe a rigorous proof is needed to say that the patterns along the diagonals are of the form $(n - a)^2$.

Why is there a perceived need to qualify the idea of proof with the adjective 'rigorous'? Julia makes the point that others have also made about prior expectations influencing our judgements about proofs. The distinction between proof and truth is significant in her account.

CLAIRE   Several people in the group tried to justify their opinions about proof—some more persuasively than others. The more convincing explanations were given by those with strong, confident voices . . . Some tried to force their view by 'intimidation'—using phrases like "surely it's obvious . . ." "it is logical to assume that . . ." which leave the listener feeling that he would appear stupid to question the explanation if he doesn't understand. Should a teacher force his opinions by means of this kind of intimidation? All these outside influences that affect our understanding and opinions make this question about proof a problem of social mathematics. Mathematics cannot be regarded in terms of black or white, i.e., understood or not understood.

The social dimension of the question was an aspect which was totally strange to all the class. The Platonist point of view had been clearly fostered very strongly throughout their mathematical training, as I suspect is almost universally the case.

CAROL   Does the above argument (that associated with Figure 2) constitute a proof? A mathematical proof is said to be a chain of reasoning in which a statement or theorem is shown to follow from previously proved statements or from initial assumptions. It certainly appears as though our deduction is valid. But is our original statement true? The relationship noticed is true on the table (Figure 1), even extending the table it still holds true but is there not a possibility that at some point it may break down? The final statement (*) has not been proved by induction since each line in our table (Figure 2) does not imply the next one. It seems to me that the problem lies in whether we can jump to our final statement (*) involving "a" from the three above. Surely they only show what seems to be a connection between the numbers in different bases. Our digits in base n have been noticed from the table which we can't do for "a" of higher values. I don't believe, in its present form, we have a valid proof.

There is a valiant attempt to define a proof here, but it is full of the unstated assumptions behind the language used that always appears in this kind of writing. It seems to me to reflect a serious gap in the mathematical education we offer to our students. Only in this kind of written assignment does this become clear.

CHRIS   The line (*) in itself is true and is a proof, provided the conditions $(n > a^2)$ are obeyed. It is an explanation of the table. The question "What is proof?" is in itself difficult to answer. It tends to come down to what you believe is true, and can be taken for granted . . . it seems to me to imply that for a proof it is only necessary to explain the "non-obvious." In this case I believe we have explained that.

This summarises the confused use of the terms "proof," "true," "explanation," and "belief" which are present, explicitly or implicitly, in the students' writing, and which were certainly there in the discussion which preceded these assignments.

Conclusions

What did the students gain from this assignment? I hope the extracts make some of that clear. They expressed clear surprise that such a variety of views about such a fundamental matter should exist among them (and were surprised when the contrast between [2] and [3] was subsequently pointed out). It made them question their own assumptions, and attempt to articulate them. In seeking to develop their skills in communication, an assignment arising in this way will have the additional driving force of the perceived need to convince their fellows. They were writing partly for themselves and for each other, and not just for an authority who would judge their work according to a pre-determined standard.

What do *we* gain from such assignments? They provide us with a window into the students' real perceptions on what we have been teaching them. The discussion in [5] is a case in point, where the points of view expressed have been unaffected by the work done in analysis about convergence of sequences and series. In the present case it made it abundantly clear that over the years we had been showing them proofs—in large quantities—without imparting any general idea about the processes of proving at all. The "ritual" of [3] applies to much of their mathematics, if the evidence of assignments of this kind is to be believed.

What do *we* gain from such assignments? They provide us with a window into the students' real perceptions on what we have been teaching them.

Although not the prime consideration, the topic of proof is itself important for students of mathematics. It is not just of interest to professional philosophers, but to mathematicians currently. Moreover, the variety of points of view expressed by the students are reflected in two well-known books. In [2], Mac Lane expresses the formality of proof, writing "Mathematics, through the decisive notion of formal proof, can appeal to an absolute standard of rigor. Moreover, this standard is external and impersonal." This contrasts with the point of view expressed by Davis and Hersh in [3], throughout which the multi-faceted nature of proof is emphasised, and the social dimension clearly acknowledged. Mathematics as "fallible, corrigible and meaningful," as "one of the humanities" encapsulates this view. The high regard in which the work of Lakatos [4] is currently held reinforces the idea of proof as an evolving, negotiated construct.

For me the important thing about the written assignment and the responses it generated was that it arose out of a lively seminar discussion in which the students expressed a clear commitment to a point of view. There are any number of situations which will give rise to this and thereby produce written work expressing the students' own ideas and thoughts as distinct from the somewhat derivative and stilted essays one sometimes encounters. The idea of mathematical induction has sparked off this kind of activity with my students, as have issues arising from the controversy about whether the decimal point nine recurring is equal to unity (see [5] for an account of a class discussion on this issue). One has to realise that the choice of topic for this kind of work will vary according to the makeup of each particular class of students, and being alive to this is vital if one is to capitalise on it in a way which I hope the extracts above demonstrate.

Finally, for intending schoolteachers in particular, assignments arising out of seminar discussions such as this force them to acknowledge the social and interpersonal dimensions of mathematics teaching and learning. This is vital for their professional futures.

REFERENCES

1. K. E. Hirst, "Undergraduate Investigations in Mathematics," *Educational Studies in Mathematics* 12 (1981): 373–387.

2. S. Mac Lane, *Mathematics: Form and Function,* Springer-Verlag, New York, 1986.

3. P. J. Davis and R. Hersh, *The Mathematical Experience,* Birkhäuser Boston, Cambridge, Massachusetts, 1980.

4. I. Lakatos, *Proofs and Refutations,* Cornell University Press, Ithaca, New York, 1976.

5. K. E. Hirst, "Exploring Number: Point Nine Recurring" *Mathematics Teaching* 115 (1986): 12–13.

Using Writing to Improve Student Learning of Statistics

Robert W. Hayden
Plymouth State College, Plymouth, New Hampshire

Let me tell you about the experience that first showed me the need for student writing in applied statistics. I had set an examination question that required my students to do a hypothesis test. It ended with a poorly worded question that students interpreted in a variety of ways. Some simply provided the results of their calculations along with a number they had extracted from a statistical table. Others included some jargon about "rejecting the null hypothesis" while others stated a conclusion in more practical terms, such as "the tested drug is probably more effective than the standard treatment." Some students provided two or even all three of these responses. Of course, all three constitute restatements of a single fact in different language. Unfortunately, I found little or no correlation between the different answers of students who gave multiple answers. If the numbers clearly indicated that the null hypothesis should be rejected or the treatment declared effective, students were just as likely to say the opposite, even when their computations were perfectly correct.

Reflecting on my students' answers, I reached a number of conclusions.

1. Since their final conclusions were no better than what they might have reached via a simple coin toss, all the complex computations I had taught them were of no real value.

2. My students' lack of understanding was mostly no fault of their own. Their textbook spent pages and pages showing them worked examples of how to do the computations, but far less space discussing what the computations meant. Exercises asked them to perform computations, but rarely asked them to explain their results. Rarely were they required to *select* an appropriate technique. The appropriate technique was always whatever technique was described last.

This led to some serious thought about what my students needed to learn in a statistics course, and how I might help them to learn those things.

I next asked myself what my students were likely to need to do with statistics after graduation. I tried to order these needs on the basis of how many of my students might have them. I hope you will pardon my listing those needs here, because they are relevant to all kinds of "book learning."

1. Virtually all of my students would need to evaluate quantitative information presented to them in newspapers, at zoning board meetings, by their doctor, or by numerous other sources. These students need to know what a median or a standard deviation is or means. They need to know the strengths and weaknesses of these numbers as summaries. They need a healthy scepticism toward quantitative claims.

2. A smaller group of my students would need to evaluate the meaning and propriety of more technical statistical techniques that might be used by researchers in their own field.

3. A still smaller group of my students might need to evaluate statistical work done by subordinates or provided by consultants.

4. A very small group of my students might actually carry out a statistical study themselves. They would almost certainly use a computer to carry out the mechanics of data storage, editing, and analysis.

5. An even smaller number of my students might one day need to carry out a large scale statistical study while stranded on a desert island, or at a remote wilderness location, or in some other situation in which a computer would be unavailable. These students would need to know how to perform the computations by hand.

If we look at most statistics books, and most statistics courses, we find them organized as if my last group of students were the norm. Indeed, the whole pyramid is inverted. Few textbook problems deal with meaning or interpretation, something *everyone* needs to know about. Instead, most deal with computational techniques that few students will ever use outside the classroom.

So, I resolved to try to spend more time on meaning, evaluation, and interpretation. However, my new found idealism was tempered by a basic fact of schooling: the students won't learn anything that does not appear on the exams. The simple conclusion is that questions involving meaning, evaluation, and interpretation must appear on the exams. Once we reach this conclusion, the need for writing is obvious: *the answers to questions of meaning, evaluation, and interpretation are verbal, not numeric.* Thus writing becomes, not just another subject to teach, nor even a tool for achieving traditional goals, but rather a necessary path to developing higher level quantitative skills.

These, then, are the values and experiences that have shaped my interest in Writing Across the Curriculum. Let me now deal with some of the practical problems of implementation. The most important piece of advice is: **start slow**. Your students have had an average of 14 years of experience with teachers who preached the importance of higher level skills but tested only on memorization and manipulative skills. Your best sermons will therefore have no effect, and your students will all fail that first exam when you ask them questions exercising skills they have never developed. You will become discouraged, curse their stupidity and your own idealism, and return to rote drill. Actually your students can do far more than you imagine, but they need your help. There follows some advice on providing that help. Bear in mind that it is based on all of the above. If your reasons for using writing assignments differ from mine, you may prefer a different approach.

The first thing you need to change is your teaching. Deemphasize mechanics. Assign only enough computational problems to get the ideas across. Keep the numbers *very* simple. Encourage the use of calculators or computers for any computations beyond the bare minimum needed to grasp the concepts. Spend lots of class time on interpretation and meaning. *Provide sample test questions!* These communicate the nature of your expectations and the fact that you are not kidding. Once you have taught the course this way a few times, you will have a bank of old exams. Share them freely. Let students see for themselves that you really do ask embarrassing questions on exams. Distribute these old exams well in advance. Students can not change their study habits the night before an exam. Indeed, you will find that they will initially, but very strongly, resist changing their study habits at all. There really is not much you can do about that except to fail those who do not perform at the level you desire. Things will improve as word gets around and students enter your class with expectations already tempered by your reputation.

Then there is the matter of writing exam questions. Start small. Problem 1 on Exam 1 should not be

> Compare and contrast the methods, assumptions, uses, and histories of parametric and nonparametric statistical techniques, giving special attention to their impact on the methodology of the social sciences.

A more reasonable start might be

> For the data 3, 1, 4, 1, 21, find the mean, mode, and median. Which of these would best summarize this data? Why?

Since many of my readers may not teach statistics, I do not want to give further statistical examples here. The principles should be clear.

Keep in mind that the main goal is to force the students to *think*. Forcing them to write is just a tool, a way to hold them accountable for thought. You do not have to make them write a lot of words as long as you get them to think a lot of thoughts. One-sentence answers may meet your goals. Also keep in mind that reading and writing may often be interchanged. Instead of asking students

> Find the slope in $y = 2x + 3$.

Or even

> Interpret the slope in $y = 2x + 3$.

You might ask

> How much does y change for a unit increase in x when $y = 2x + 3$?

Now the answer is a single number—much easier to grade than a student-written sentence or paragraph on the subject.

Sometimes teachers are discouraged by the quality of writing they get, or discouraged from asking for writing by fears of what they *might* get. In my experience, lack of mastery of subject matter will far outweigh any writing flaws. Indeed, you may discover that your students know far less than you thought about the meaning of those numbers you taught them to calculate. This can be taken as a sign of either the futility or the importance of your work, depending on your outlook on life. You should work on teaching your discipline until the content of the answers is better than the expression. In the process, you will find that the expression improves by itself. No one communicates well when they have not the faintest idea what they are talking about.

Sometimes teachers are discouraged by the quality of writing they get, or discouraged from asking for writing by fears of what they *might* get.

Yet another issue is grading student writing. Here my solution is as simple as it is radical: *don't*. I grade students only on such knowledge of statistics as they are able to communicate to me. As long as their mastery of the mechanics is good enough so I can understand what they are saying, they can get full credit. The only grammatical advice I ever give is, "Never start your first sentence with a pronoun." Many of my students are as anxious about grammar and punctuation as they are about mathematics and statistics. For better or worse, I try to handle things so they never notice the writing in the course. My exams are what they are to reflect what statistics is about, not to reflect what writing is about.

However, there are some things on the border line between statistics and rhetoric that I do take into account. I prefer short, direct answers. (Often students are amazed at how short an answer I will

accept.) Ambiguity or vagueness is taken as a sign of uncertainty and costs points. So do irrelevancies. I insist that students read the question carefully and stick to it. Indeed, the biggest problem I find (other than lack of knowledge of statistics) is failure to answer the question asked. This, of course, is a problem of thought rather than syntax.

I have been writing as if all the writing I require is on exams. That is very nearly true. Remember that I am trying to find ways to get students to think and ways to hold them accountable for thinking, and exams are the ultimate accountant. I have experimented with projects where students analyze a set of data and write up a report, but I have had less success with this. Just worrying about what the numbers mean is a wrenching change for many students. Asking them to consider the meaning of dozens of numbers and integrate them into a report is really too much to ask at the beginning. Perhaps this will change as other instructors, especially those in the high schools and grade schools, start to emphasize meaning and interpretation.

Another facilitator of change would be a better selection of textbooks. I mentioned earlier the reciprocal relation between reading and writing. Students are more likely to write well if their own textbook does so. Of course, the text should also support your emphasis on understanding and interpretation, while not making unrealistic assumptions about your students algebraic skills or reading level. I find Siegel's introductory statistics text to be excellent from this point of view.

In an age of assessment and accountability, perhaps I should close with some sort of "evaluation" of the success of what I have been doing. This is impossible. I have no idea of what students thought a standard deviation meant before I started asking them. Based on their answers during the brief transition period, before they expected such questions on exams, my suspicion is that it never dawned on them that a standard deviation *had* a meaning. It was just a cue-word used to Pavlovically stimulate a certain computation. On the other hand, I have often noticed that mathematicians and statisticians are among those *least* compelled to quantify everything, perhaps precisely because they *do* know the meanings of numbers—which entails knowing which numbers are meaningless. For me it is enough that today much of my students' attention is directed toward the parts of statistics that I consider most worth knowing. A decade ago almost all their attention was devoted to the parts least worth knowing. I cannot quantify that change, but I can tell you it is a very important change, and a change that could only have been brought about by making students write.

BIBLIOGRAPHY

P. Connolly and T. Vilardi, *Writing to Learn Mathematics and Science*, Teachers College Press, New York, 1989.

A. F. Siegel, *Statistics and Data Analysis: An Introduction*, John Wiley, New York, 1988.

W. Zinsser, *Writing to Learn*, Harper & Row, New York, 1988.

Integrating Writing into the History of Mathematics

Dorothy Goldberg
Kean College of New Jersey, Union, New Jersey

It is generally agreed that large numbers of college students need to improve their writing skills. Surely, students learn to write by writing, and they should write often, not just in their English courses, but across the entire college curriculum [1].

At Kean College of New Jersey, one of eight state colleges, the faculty is fully committed to providing opportunities for students to develop their writing competency beyond just the first two undergraduate years of general education.

Surely, students learn to write by writing, and they should write often, not just in their English courses, but across the entire college curriculum.

Faculty members in every department have developed writing-emphasis (W-E) courses, which are subject matter courses that differ from other courses in that their syllabi call for regular writing assignments and their instructors are committed to writing as a means of mastering course content.

Official college policy requires that all students complete two W-E courses in their junior and senior years: one W-E course in the major program and the other outside the major.

The administration is also committed to the writing-emphasis program. Several workshops and retreats have been held to encourage and assist faculty in developing their W-E courses. In addition, extra financial compensation has been provided.

The procedure for certifying W-E courses is simple. In the case of already existing courses modified for writing-emphasis, the department curriculum committee must approve the W-E course and then forward the course description to the College Writing-Emphasis Committee for approval. In the rare cases where an entirely new course must be designed to accommodate writing-emphasis, the new course must follow existing procedures for course certification: departments forward course descriptions to school and college curriculum committees before they send them to the Writing-Emphasis Committee.

The College Writing-Emphasis Committee provides criteria and format requirements for W-E courses.

CRITERIA

1. Each W-E course should include a variety of regular writing assignments. Instructors should make requirements and standards for evaluation of each writing assignment explicit to their students.

2. Each W-E course should include some form of writing (graded and/or ungraded) every other week (in a 15-week semester).

3. Instructors should respond in a timely fashion to writing assignments submitted by their students.

4. Each W-E course should include a core of three to five graded writing assignments that follow a sequence.

5. It is understood that the Writing-Emphasis Committee, like any other college wide committee, will use discretion in applying these criteria to various programs within the spirit of the college policy on W-E.

FORMAT

The following four items should be appended to existing course descriptions:

1. Types of writing: a list of the types of writing assignments required;

2. Rationale: an explanation of the appropriateness of types of writing assignments for specific subject matter treated in the course;

3. Methods of evaluation: a statement of the proportion and frequency of graded and ungraded writing; the extent of the course grade based on writing; the criteria for evaluating writing; the method of written response to student writing; and

4. A sample sequence of writing assignments: a fully explicit model of three to five writing assignments that follow a sequence, making clear to students the procedures required and criteria for evaluation [2].

The History of Mathematics W-E Course

The criteria and format as specified by the Writing-Emphasis Committee were satisfied in developing the History of Mathematics as a W-E course.

A sample sequence of writing assignments was submitted:

1. *Rationale for choosing a mathematician about whom an oral report is to be given*, assigned at first class meeting, due second week of semester. The first writing assignment gives students an opportunity to express themselves in a simple format in a natural way.

TO STUDENTS   Read the directions for *Report on a Mathematician* (see Appendix A). Next week please submit the following information:

A. Your name

B. "Your mathematician"

    1.
    2.
    3.

You are asked to make three choices so that duplicate reports can be eliminated and you can still have some freedom of choice.

C. Write an essay of at most one page in which you give your rationale for choosing "your mathematician." Do not give as a reason that the choice was based on whether the report would be given early, late, or in the middle of the semester. Be clear and concise. Be specific. Use simple language.

2. *Analytic essay on journal article or film,* assigned at first class meeting, due 5th, 8th, 11th, and 14th weeks of semester.

In these assignments the students are given the opportunity to write about readings and films that enrich the course beyond the reading of the text. They are asked to determine the central theme or focus of the film or article and to describe what new ideas they learned and what was enjoyable about the assignment (see Appendix B).

On the day the assignment is due, students exchange papers in pairs and make editorial suggestions to each other. The original paper is collected by the instructor. At the next class meeting a second draft is turned in to the instructor.

3. *Essays on Examinations.* Both the midterm and the final examinations contain some short essay questions.

    The midterm and final exams cover the text readings and exercises and the class discussions and reports.

4. *Term report (research paper).* The term report is due at the end of the semester. Students are required to write this research paper or to take the final examination (which contains some short essay questions). For directions to the students, see Appendix C.

How is the writing evaluated?

Except for the first assignment, all writing assignments are graded. The criteria for evaluation, as defined for the students, are clarity and conciseness, completeness and significance, and ultimately style.

Whether or not the students edit each others papers first, the instructor makes comments on the students' papers. These comments may be corrections, suggestions, and praise.

As for computing the grade for the course, a point system is used. If one-third of the midterm and final exams consists of essay questions, then depending on whether the students take the final examination or write the term report, 38.5% or 55% of the grade is for writing (see Appendix D).

As described in this paper, the History of Mathematics was taught as a writing-emphasis course during the Spring 1987 semester at Kean College of New Jersey.

As measured by the commitment, excellent participation, and cooperation of the students and the outstanding educational results, it was a very satisfying course, for students and instructor alike.

REFERENCES

1. Dorothy Goldberg, "Integrating writing into the mathematics curriculum," *Two-Year College Mathematics Journal*, 14 (November 1983): 421–424.

2. Richard Katz, "Writing emphasis course descriptions: criteria and format," Kean College of New Jersey Writing Emphasis Committee, 1987.

Appendix A

Report on a Mathematician   One of your assignments is to report to the class on the life and work of a particular mathematician.

The purpose of this assignment is to give you experience in using our library for research and to provide a change of pace during our class meetings. Each member of the class will act as guest lecturer!

On the assignment sheet that you will be given, some suggestions are made about how to start your work.

When you are asked to look up works by and about the mathematician assigned, use 4" by 6" index cards to record the information you need (or slips of paper). Each should contain the author's name, title, date of publication, publisher and place of publication, the library call number and any other information that is of interest to you. Use the following format:

HARDY, G.H.
1967   <u>A mathematician's apology.</u>
          London: Cambridge University Press.
QA7   Autobiography of a famous English number theorist.
H3   Interesting biographical introduction by C. P. Snow

Your report should take 20 minutes. Include information about "your mathematician's" life, contributions to mathematics and other fields, whatever you have found of interest, e.g., anecdotes. You must explain some mathematical contribution by "your mathematician." If you find that you'll need more time, please consult me.

You must supply references you have used on your index cards, since part of the intent of the assignment is for you to read a variety of materials about or by your subject.

You will be the class expert on "your mathematician," so be prepared to add to discussions whenever that person's name is mentioned.

Report on Mathematician

Dr. D. Goldberg

Your Mathematician (YM) is _____ .

0. Look up YM in Burton's text. Read the relevant pages and take notes.

1. Look up YM in the *Dictionary of Scientific Biography* (ref. Q141. D5).

    Make an index card which includes the following information:

    > The article is in volume _____ , pages _____ , and is by _____ . You should read the article and take notes.

2. Look up YM in an encyclopedia. On an index card give the name of the encyclopedia, date of publication, volume, pages, and author (signed articles are more reliable). Read the article and take notes.

3. What works *by* YM are in the KCNJ library? (You need only make index cards for each publication, including a brief description of the contents).

    A typical index card is as follows:

BOOLE, GEORGE
1958   <u>An investigation of the laws of thought.</u>
          New York: Dover
BC135   Boole's work on logic and probability and statistics.
B7   Not easy to read. Originally published in 1854.

4. What works *about* YM are in the library? Read any that interest you and take notes, if you need them. Make index cards.

    A typical index card is as follows:

CARDANO, GIROLAMO
1953   Ore, Oystein. *Cardano, the gambling scholar.*
        New York: Dover.
Q143   A wonderful biography of Cardan, with good
C307   mathematical explanations. Easy to read.

5. Try to find some anecdotes about YM. Check Eve's *Mathematical Circles* books (QA99, E83 or QA99, E842). Make index cards.

6. There may be journal articles about YM. Check the following indices: *Biography Index, Social Science Index, Education Index, General Science Index, Essay and General Literature Index.* It is also possible to check the cumulative index of a journal most likely to contain articles on the history of mathematics: *Historia Mathematica, Isis, Scientific American, Mathematics Teacher, American Mathematical Monthly, College Mathematics Journal, School Science and Mathematics.*

You should have at least one journal reference on an index card.

A typical index card is as follows:

LOVELACE, AUGUSTA ADA BYRON

    Kean, D.W. "The computer and the countess." *Datamation* 19 (May 1973): 60–63.

    A great little article about our first computer programmer. Fascinating to read.

Note: When you give your talk, please turn in your reference cards. I will keep them.

Appendix B

Writing Assignments
In addition to the term paper (or to the final examination) there are four short writing assignments in this course.

To fulfill this requirement, you may either view a film at the Instructional Resource Center (J230) or read a journal article:

Examples of Films

M2-840   *Numbers Now and Then*
(Babylonian, Egyptian and Greek numerals)   (25min.)

M2-733   *Shaking the Foundations*
(Set theory: 19th century dilemmas)   (25 min.)

M2-734   *A Time for Change*
(Early history of calculus)   (25 min.)

M2-233   *Music of the Spheres*
(52 min.)

Journal Articles.
There are two journal articles on Newton specifically mentioned on the last page of the bibliography.

For other journal articles, see the directions given on page 2 of "Report on a Mathematician."

Two of my favorite articles are:

    D. W. Kean, "The computer and the countess," *Datamation*, 19 (May 1973): 60–63.

    D. W. Kean, "The author of the analytic engine," *Datamation*, 12 (January 1966): 37–42.

Directions. Be sure to choose a film to see or an article to read that you truly enjoy spending time on. It's much easier to write about something that was stimulating, interesting or enlightening to you.

Be clear and concise in your writing. Use simple language. Write as though you are communicating with another student in the class. Include answers to the following questions.

What was the central theme or focus of the film or article?

What did you learn that you had not already read about or studied in class?

What did you enjoy about the film or article?

Two typewritten pages are sufficient. Bring in your original paper plus an Xerox copy. Also, if you read a journal article, turn in your Xerox copy of the article with your paper.

Appendix C

Term Report
You are to write a paper on a topic of your choice. This is meant to be an interesting and enjoyable assignment, not a chore. So choose a topic with care. The only restriction I will impose is that it cannot be on the mathematician that your first report dealt with.

Some of the categories your topic might come from are:

1. Concept evolution, e.g., the development of the function concept.

2. Topical subject, e.g., history of quadratic equations.

3. Biographical, e.g., life and work of a particular mathematician.

4. Cultural and philosophical impact of some idea in mathematics, e.g., perspective, non-Euclidean geometry.

5. Focus on a portion of some era in mathematics, e.g., Hindu, African, Chinese, Japanese, Hebrew.

Each paper must meet the following requirements:

1. The paper is to be on the history of mathematics. It cannot be all history or all mathematics. It should contain a reasonably nontrivial piece of mathematics as well as the history and background of that mathematics.

2. Enough expository material should be included so as to make the paper self-contained.

3. You should use a variety of research materials and must give careful references to your sources. You will want to use books and encyclopedias, but I especially encourage you to use the journals. Your paper should include a bibliography listing your sources and they should be cited in the body of your paper when appropriate.

4. I strongly prefer that the paper be typed, but the length, format, etc., is up to you. You should make a carbon copy or Xerox copy since I intend to keep one copy of your paper.

The grading of your paper will be based on a number of factors, including: the historical and mathematical content; the significance, interest, accuracy, and completeness of the material; the scope and significance of your references, and the sensitivity with which they are used and cited; the style in which it is written. Poorly written papers will not be accepted. The grade of A will be given only for truly excellent work; B for good solid work; C for acceptable work; D or F for unsatisfactory work. All grades are possible.

To make certain that you devote sufficient thought to your paper, begin by selecting your topic. The due date is April 6th. A report is to be submitted which should include (a) your topic, (b) a few words about what you intend to do, (c) your preliminary bibliography, and (d) any questions you have about your paper. The intent of the preliminary report is twofold. It should aid you in taking the writing of your paper seriously. It allows me to make suggestions.

The final version of your paper is due on May 11th.

For suggestions about writing your paper, read pp. 1–33 of Kenneth May's book and look at the topics listed in the bibliography.

Appendix D

GRADING
Your grade for the semester will be based on your assignments, test scores, and your contribution to class discussion.

Midterm exam	60
Report on a mathematician	40
Four short writing assignments	40
Term paper or final examination	50
Contribution to class discussion	10
	200

TESTS
The midterm examination will be given on _____ .

The final examination will be given on _____ .

A sample exam will be distributed before the midterm exam. There will be true-false questions, fill-in-the-blanks, matching questions, short answer questions, and essay questions. The exams will cover the text readings and exercises plus the class discussions and reports.

SUMMARY OF DATES

_____	Choosing a mathematician
_____	Writing assignment 1
_____	Writing assignment 2
_____	Midterm examination
_____	Writing assignment 3
_____	Term paper topic with partial bibliography
_____	Writing assignment 4
_____	Term paper
_____	Final examination

Writing to Learn and Communicate Mathematical Ideas: An Assignment in Abstract Algebra

Anne E. Brown
Saint Mary's College, Notre Dame, Indiana

In designing writing assignments to enhance learning in mathematics majors' courses, I set two principal goals:

1. The assignment must be an integral part of the coursework.

2. The assignment must require the student to make a significant effort to communicate the results in writing.

An assignment that I have found to be consistent with these goals is one in which the student writes an account of the proof of a specified theorem, and includes the necessary definitions, examples, lemmas, and expository remarks leading up to the proof. Ideally, through the process of writing, refining and rewriting, the writer develops a clear overview of the topic as well as a local grasp of the details, and learns to communicate mathematical ideas accurately and effectively.

This essay describes the structure and evaluation of one of two writing assignments that I use in my junior level abstract algebra sequence. These writing assignments feed into an advanced writing program that requires each mathematics major at Saint Mary's College to assemble a portfolio of three acceptable papers in mathematics: one from a sophomore level course, one from a junior level course, and a senior paper that is a report on a major independent study project. In the sophomore year, each student receives a pamphlet detailing the departmental guidelines used for evaluating writing in mathematics. She is expected to strive to meet the standards in all written work, including daily assignments as well as formal writing assignments. (Interested readers may contact the author for a copy of the guidelines, and see reference [1] for a description of the advanced writing program in place at Saint Mary's College.)

A Sample Writing Assignment

The students in my abstract algebra sequence complete two writing assignments each year: one covering a topic in elementary group theory, and one covering a topic in ring or field theory. The mathematical goal of the first assignment is to prove that every infinite group has infinitely many subgroups. To provide a focus for the discussion that follows, here is a brief sketch of a proof:

THEOREM. *Every infinite group has infinitely many subgroups.*

Sketch of proof: Let G be an infinite group. There are two possibilities: either G contains an element of infinite order, or G contains only elements of finite order. If G contains an element g of infinite order, then $\{\langle g^i \rangle : i \in Z, i \geq 0\}$ is an infinite collection of distinct cyclic subgroups of G.

On the other hand, suppose that G contains no elements of infinite order. If G has only finitely many subgroups, then G has only a finite number of distinct cyclic subgroups. But every group is the union of its distinct cyclic subgroups, and hence G is a finite union of finite sets. This is a contradiction, so G must contain infinitely many subgroups.

Despite the fact that the proof of the theorem involves only elementary concepts of group theory, my students have found that writing a complete account of its presentation is a challenging assignment. Indeed, one reason for choosing this particular theorem is that the student is forced to organize, summarize, and illustrate most of the basic concepts covered early in the course.

The audience for the paper is assumed to consist of mathematics students with some knowledge of abstract algebra, but who need to be refreshed on the fine points of elementary group theory. Insisting that she write for an audience other than the course instructor helps eliminate the student writer's tendency to rely on the reader to correctly interpret incomplete or unclear explanations. This is practical preparation for writing outside the classroom, where she would likely be expected to communicate technical results to those with less expertise in her area. The student writer thus becomes aware that the issue is not only whether she understands the topic, but also whether she has found an effective way to explain it to a specific audience.

> **Insisting that she write for an audience other than the course instructor helps eliminate the student writer's tendency to rely on the reader to correctly interpret incomplete or unclear explanations.**

Since the goal is to combine good writing with correct and complete exposition and proof, I give the students as much help as they need in organizing the material, especially for the first writing assignment. Most junior math majors are inexperienced in assembling a coherent presentation of the definitions, examples and results needed to fully discuss and explain the proof of a theorem, so I provide them with an outline. Generally, the components that must be included are:

1. An introduction identifying the major point of the discourse, and expository remarks indicating the writer's plan of action.

2. Definitions, preliminary results, and examples: order of a group, order of an element, cyclic subgroups, connection between the order of an element and the order of the cyclic subgroup it generates.

3. A lemma characterizing the distinct subgroups of the cyclic group $\langle g \rangle$, where g is a group element of infinite order. (See exercise 2, below.)

4. Example: an infinite collection of subgroups of the group of integers **Z**. (This might appear before item 3, if the writer wishes to motivate the lemma.)

5. Proof of the theorem: Every infinite group has infinitely many subgroups.

6. Example: an infinite group containing only elements of finite order, and an infinite collection of subgroups of the group. (See exercise 3, below.)

7. A conclusion.

The first drafts tend to be much better if some of the work indicated above is completed before the formal writing assignment is given. This is achieved by including the following problems as part of regular homework assignments given earlier in the semester:

1. Write a 1–2 page essay explaining the concepts of the order of a group element, cyclic subgroups, and the relationship between the order of elements and the order of cyclic subgroups.

Include definitions and examples. Explain the notational differences when the group operation is addition, rather than multiplication. Explain why every group is the union of its distinct cyclic subgroups.

2.

   a.   Prove: If g is an element of infinite order in a group, and k and j are nonnegative integers such that $\langle g^k \rangle = \langle g^j \rangle$, then $k = j$.

   b.   Use the result in part (a) to identify an infinite collection of distinct cyclic subgroups of the group of integers **Z**.

3.   Let R_α denote a rotation of the plane a degrees counterclockwise about the origin. Let $G = \{R_\alpha : 0 \leq \alpha < 360, \; \alpha \text{ rational}\}$.

   a.   Show that G is an infinite group under composition in which every element has finite order. (Note: $R_\alpha \circ R_\beta = R_\lambda$, where $\lambda = \alpha + \beta \pmod{360}$.)

   b.   Show that $\{\langle R_{360/n} \rangle : n \geq 2\}$ is an infinite collection of distinct subgroups of G.

Examples and Student Writing

The mathematically mature student should be able to state a definition or theorem and apply it in a specific instance; he or she should also be able to formulate conjectures about general results based on examples. Demanding clear and carefully written analyses of examples in writing assignments is one way to help develop these skills.

The most common problem I have seen with examples is that students will supply the computational details, but will fail to point out the significance of the example, expecting the reader to know why it was included and what it illustrates. I remind them that the rules of mathematical exposition are the same here: they should indicate the purpose of the example, give the details, and then summarize their conclusions in the context of the definitions or theorems under discussion. To help them clarify their explanations, I point out that examples are often used to illustrate a point under discussion, to show a specific instance that suggests a general theorem, or as motivation that leads the study in a new direction.

Discussion and Student Writing

Some students need more help than the outline provides in order to be able to write proofs of the results and then tie the results together with insightful expository remarks. My strategy is to avoid writing any more of the details for them, but rather to direct informal discussions (outside of class) where they offer ideas and we, as a group, refine them verbally. Those present take notes on the discussion, and usually manage to write coherent proofs in rough drafts that are later revised and clarified. This format has the added benefit of encouraging collaboration and discussion about mathematics among the students.

Student writers usually have some difficulty writing the introduction and the necessary expository remarks because they don't see the "big picture." This is not surprising, since a typical homework assignment consists of working unrelated exercises, where a unifying conclusion is rarely called for. One purpose of the discussion session is to help them develop a broad overview of the subject matter,

so that they can write confidently about the preliminaries, with the final goal always in mind. For instance, I ask them to explain how the lemmas and examples relate to the proof of the main theorem. In particular, talking about the differences between the two examples (in parts 4 and 6 of the outline) helps them to identify the important concepts and to see why the proof of the main theorem falls naturally into two cases.

Evaluation

There is little doubt that reading and evaluating writing assignments is an added burden on the teacher. Here are some ways we at Saint Mary's have found to make the task easier:

1.   Assign some of the writing as part of the regular homework. For best results, the teacher (rather than a student grader) should read and comment on the solutions or essays.

2.   Give a detailed outline of the assignment, and make your expectations clear.

3.   Do not spend an excessive amount of time commenting on first drafts. It is unlikely that the student will read or understand all of your comments, so try to focus on the major errors. This also helps avoid drowning the less successful student in a sea of red ink. For most student writers, you might as well wait until a later draft to give advice about how the overall exposition could be improved.

4.   Return the papers promptly, and require submission of the next draft as soon as possible, while their ideas are still fresh.

5.   Emphasize good writing on all homework assignments. Sometimes I return a correct but poorly written homework solution to a student with instructions to revise and rewrite it. Credit on the entire homework assignment is withheld until the revision is submitted.

6.   Meet informally with small groups of students to discuss specific mathematical problems with the assignment. This should ensure that most of the mathematical errors are corrected before you see the first draft.

7.   When appropriate, encourage the use of a word processor to facilitate quick and easy revisions. At Saint Mary's, we require that a word processor be used on the papers that are intended for the student's portfolio.

Whether our students become professional mathematicians or not, it is likely they will find that writing, revising, and rewriting is required in their careers.

Conclusion

I believe that gaining writing experience in mathematics courses is essential to the serious study of mathematics, partly because working mathematicians write, refine, and rewrite as part of the creative process. In writing to communicate mathematical results, the discussion of preliminaries might change repeatedly as the researcher simplifies a long or complicated proof. In this spirit, I have long urged my students to write rough drafts of problem solutions, and to refine and rewrite their solutions before submitting them. Before making formal writing assignments, I never saw evidence that they wrote any drafts at all. I still cannot claim that all (or even most) of my students write drafts routinely, but I have seen great improvement in the clarity of the homework papers of many of my students. Writing drafts may also broaden the student's view of how mathematics develops. One of my algebra students commented that she had always felt that proofs of theorems must emerge complete from the mathematician's mind. Through working on the writing assignments, however, she came to believe that the creation of new mathematics really is a process. It is interesting that her own experiences, rather than class lectures or textbook presentations, convinced her of that fact.

Whether our students become professional mathematicians or not, it is likely they will find that writing, revising, and rewriting is required in their careers. Many of the graduates in mathematics from Saint Mary's obtain positions in business or industry. They report that they spend a significant amount of time writing technical memos and project documentation, and that their college experiences in writing mathematics (particularly those involving organizing, summarizing, and exposition) were excellent preparation. This information reinforces our belief that we are on the right track in requiring writing, and it is influential in motivating our students to work to improve their mathematical writing.

REFERENCE

1. Joanne Erdman Snow, "The Advanced Writing Requirement at Saint Mary's College," in Paul Connolly and Teresa Vilardi (eds.), *Writing to Learn Mathematics and Science*, Teachers College Press, Columbia University, 1989, Chapter 15.

Writing in a Non-Euclidean Geometry Course

Richard S. Millman
Wright State University, Dayton, Ohio

1. Introduction

Writing across the curriculum (WAC) has become extremely fashionable these days. Most universities either have a program for WAC in place or are forming a committee to make recommendations about it. The impetus for many of the writing programs comes from people in industry who complain that our students cannot write. Those of us in the mathematical sciences are all too aware of the difficulty in reading technical manuals for computer or word processing software. However, we are not without sin ourselves. Indeed there is a remark which I have heard attributed to the Nobel laureate C. N. Yang to the effect that mathematics books are written in two ways—those that are so poorly written that the reader cannot make it through the first paragraph and those that are so poorly written that the reader cannot make it through the first sentence.

Frequently, science and mathematics students have no idea that they will be required to write when they leave school and, therefore, are not at all sympathetic to writing assignments. The situation is further complicated by the fact that many faculty members in the sciences are not aware of the philosophy of WAC. In fact, many science departments that require a laboratory feel that the lab reports themselves are enough of a writing exercise. This belief misses the point of writing intensive projects, which is that there must be multiple drafts. The point is, (Young and Fulwiler [7]),

> The teacher who examines what happens during the whole process, in addition to what is demonstrated by the final product, learns where students are having problems and where and what kind of help they need. (p. 38)

The use of writing in mathematics courses frequently meets with faculty reservations because it is not clear how one would write in such classes. The pilot project that I carried out in the summer of 1988 was to teach a geometry course with a writing assignment. The purpose of this article is to describe the project, the student reaction to it, and both the benefits and the difficulties inherent in such an approach. The next section gives a description of the project itself and its ground rules. Evaluations were made formally by asking the students for their opinion (via a questionnaire) and informally by me from classroom and office hour contact. Those observations are described in Section 3. The consequences of this writing intensive project are discussed in Section 4 and conclusions drawn in the final section. The questionnaire, its results, and the actual topics chosen by the students are included as an appendix.

2. Project Description

First, a few words about the course and its audience. The course was Math 471, Geometry, a junior/senior level course that at many universities is described as a "geometry for future secondary teachers" course. It may also be taken for graduate credit by students who go into more depth in the various topics. A metric approach to geometry was used and the topics covered included incidence and metric geometries, betweenness, the plane separation axiom,

the properties of angle measures and the various congruence theorems. This corresponds to the first six chapters of Millman and Parker [4]. The course met three times a week for ten weeks which is the usual pace for Math 471 whether it is a summer or academic year course. The audience was a mix of sixteen undergraduates and four graduate students (whose work was more extensive than that described above). There were four mathematics majors and fifteen students majoring in secondary mathematics. The average grade point average of the undergraduate students was 3.04.

The project was a tremendous amount of fun and a marvelous learning experience for me.

The writing assignment was a multidraft project about an enrichment topic. (Appendix 2 lists the topics which were actually chosen by the students.) To make sure that there would not be just a final version turned in, a first draft was required four weeks after the class had begun with the final version due two weeks later. Although there was no way to guarantee that the students would actually hand in different papers on the two dates, that was actually the case. The guidelines for evaluation (which were distributed in advance to the class) follow.

1. All drafts were reviewed in detail and returned with copious remarks on them. They were all examined without prejudice to the final grade of the student. Students were encouraged to hand in more than the two versions required by the ground rules.

2. The papers were judged on the depth of the topic, the mathematics and the skill of the presentation, including grammar, punctuation, and spelling!

3. The intended audience for the paper needed to be clearly indicated. The students had to identify whether they were writing for K–6 students, 7–9 students, high school students, teachers at the secondary level, etc.

4. The paper had to be long enough to cover the material and show its relevance. Because this was quite subjective, I did want to give the class an idea of a reasonable length. A suggested length of five pages was chosen because it would allow someone to explore a topic without causing undue pain to those who had no interest in writing. (After all, writing was not the main part of the course.)

5. Any topic could be chosen but I did suggest the following list. Any article from *Learning and Teaching Geometry, K-12, NCTM Yearbook* [6] or *The Mathematics Teacher* 78, No. 6 (September, 1985) would be appropriate. Informal geometry (see Hoffer, [3]), van Hiele levels, the presentation of a model of a geometry as in Byrkit [1], for example, or a specific "unusual" model mentioned in the text, transformation geometry, history of geometry, computational geometry or computer graphics were topics that were also singled out as possibilities. These areas were suggested because the articles are relatively easy for those at the undergraduate level to understand; they are in our library; they deal with geometric topics; and they are of current interest in mathematics or mathematics education.

6. The final grade was determined only by the final version of the paper. It had to be labelled as final and turned in by a specific date. No late papers were accepted. The paper was weighted as one half of one of the two exams. (The homework counted as a third exam.)

3. Observations

The first observation from the point of view of the instructor is that the project was a tremendous amount of fun and a marvelous learning experience for me. The level of the student's mathematical ability is known to the professor especially after the first exam; however, the writing exercise revealed another dimension of the student's ability. Given the discussion in the press about the way that our students cannot write, I was pleasantly surprised that even the rough drafts were quite acceptable. There were certainly grammatical errors (i.e., subject-verb disagreement, sentence fragments, etc.) but the grammar was reasonable and the spelling adequate. This may be because our student body is older than the usual 18–22 year old group. The average age of the undergraduates in Math 471 was 30 with youngest being 21 and the oldest 48.

The problems with their initial drafts will be presented in order of decreasing number of occurrences. The most common difficulty with the initial draft was organization. The students rarely began their papers by saying what they were writing about, where they were going, and how they were going to get there. Most frequently, they would start the paper with what would normally be a second section—the body of the paper itself. This difficulty may well be caused by their lack of audience awareness—they are writing what is called "writer-based prose." The problem was easily remedied by either adding an introduction or by rearranging what was already present.

A second problem in the initial drafts was the frequent repetition of words and phrases. It seems that they were unaware that the English language is flexible enough to allow for non-repetitions. Certainly, this is difficult if not impossible in very technical articles, but the class would not even attempt to find synonyms for words of standard English. In fact, they were surprised that I made comments about this aspect of their writing.

The next most common difficulty was the tendency toward choppiness, even on the final drafts. They did not seem to have a feeling for what will "scan well." (All the students are native speakers of English.) Certainly, this is a subjective complaint but the writing hit my eyes wrong often enough that this was clearly a problem. I found this concept very hard to communicate to the students and will, in the future, seek help from the composition faculty to deal with this problem.

To my surprise, the students had difficulties reading the articles in *The Mathematics Teacher*. Those which dealt with the van Hiele models or informal geometry did not cause much stress, but the ones which were more mathematical did. Certainly, papers in the mathematical research journals, even *The American Mathematical Monthly* would be hard, but the level of *The Mathematics Teacher* should not be that strenuous. The last observation was that in the initial drafts from members of both sexes there was sexist language. This was easily cleared up and the suggestions for correction met with a receptive audience.

Indeed, ten of eleven students said that they learned more about geometry because of the paper than they would have without it, and nine out of ten said that writing was a valuable part of the course.

Having dealt with the viewpoint of the professor to the project, let me turn to that of the student. These responses are gleaned from the questionnaire (Appendix 1), and from more informal contact. The key point is that the students agree on the high value of the assignment, if not to geometry itself then to the greater good of education, and they unanimously found it interesting. Indeed, ten of eleven students said that they learned more about geometry because of the paper than they would have without it, and nine out of ten said that writing was a valuable part of the course. The one person who responded "no" to that last question added in her overall comments that it "was not a valuable part of the course (i.e., did not add to my knowledge of metric geometry). It was a valuable tool for broadening my outlook on mathematics—I looked beyond the 'how-to's' into the why's." This student showed a myopic view of mathematics, or at least her coursework. The writing assignment rectified that! For that student, the use of writing in the course was certainly an unexpected benefit.

A major advantage to this kind of assignment also mentioned in the survey (12-0 in favor) is that the students were forced to do some independent reading in the literature of mathematics and mathematics education. This was the first time that some of them had ever needed to do that. As most of these people will be teaching math at some level, an introduction was quite important. In fact, a number of them commented explicitly in the written comments section of the questionnaire that they were very happy to know of the existence of such journals as *The Mathematics Teacher* and *The American Mathematical Monthly*.

From the survey, it is clear that the list of suggested topics and the format (as described in section 2) were considered appropriate by the students. There are two conclusions that can be drawn from this response. First, there does not need to be a large range of suggested topics in the initial assignment. Second, students do not mind (and may actually like) the multidraft format. I was surprised that only about half of the students actually used word processing equipment. They should have learned the value of such automation by their junior year.

4. Conclusions

First, the positive consequence of writing in this geometry course will be discussed. According to the survey, the students felt that they learned the course material better than they would have without the project. I don't agree, but I do believe that they learned a lot more of what is relevant to their career goals than they would have otherwise. They were able to tailor their paper topic to their future career which for most was teaching at the junior high or high school level, and so informal geometry and taxicab geometry were popular choices. At a different end of the spectrum, the philosophical foundation of geometry was the topic of a graduate student interested in logic. After watching the drafts go by, I agree with the students that they learned something about writing in mathematics. The writing required them to work on their own, sometimes with only minimal supervision on my part. Thus, they experienced a level of independence which is not usually a part of the undergraduate experience in mathematics.

A major advantage of a writing assignment is that it enabled the class to have what is called in the assessment literature a capstone experience; i.e., a project that requires students to use a number of skills they have learned during their college career to work on a major project. The point is that they will synthesize different aspects of their education in the process of finishing the assignment. While the geometry writing project was quite small compared to capstone projects, the students did comment on the necessity to synthesize mathematical literature, composition, and mathematics. (Dare I say reading, 'riting, and 'rithmetic?) I do hope that this gave them some insight into what is expected of professionals in their careers.

The major negative consequence of the assignment (and it would apply no matter what the course was) was the large increase in workload for both the professor and the students. The class said that they averaged 18 hours each on the assignment with the high being 45 and the low 6. It took me about a half an hour to read one draft by one student. With the number of possible drafts being limitless, this could have caused a major problem. As it was, I spent about 15 hours on the first drafts, eight on the final ones and another ten hours on intermediate drafts. A 23 hour increase in workload over the course of a quarter is an extra 2.3 hours per week. Students also tended to spend more time in my office discussing their project. This added to the professorial time dedicated to this project. (I did not keep a record of this.) Since commenting on writing is not anything that should be done by GTAs, this method of instruction really adds significantly to a professor's workload. Although it is not a negative statement, I was surprised by the response to the survey question which indicated that the students did not feel that they got to know their instructor better because of the project. Perhaps the ones that would have come in for help did so anyway and those that wouldn't have, did not anyway.

Many of the experiences that I had this summer are transportable to other junior/senior level mathematics courses and other institutions. The approach is certainly not dependent on the text (Moise [5] or Greenberg [2] would certainly have worked as well). In the semester system, giving a writing assignment would probably work even better than the quarter system because there would be more time to have a mathematics background before the first draft was due.

There are a few ingredients which are crucial for such an experiment to succeed. The instructor must make a real time and energy commitment to the project. The professor needs to have a very positive attitude toward writing and serve as its advocate. There must be a reasonably broad choice of topics. Word processing equipment needs to be available and accessible on campus. Guidelines need to be decided upon and distributed in writing. Rough drafts with comments on them must be returned immediately so that the student can work on a revision.

Would I teach the course this way again? Absolutely.

If I had to do it all over again, what would I change? First of all, I would distribute the guidelines for the writing assignment earlier. It was a mistake to wait for two weeks into the quarter to do it. I would tell students in advance of the class meeting that 20 minutes of that class meeting would be devoted to an explicit discussion of what was expected. I would have the students number their drafts and keep them in a folder which they would turn in each time. I would warn students in advance that their papers will come back with many marks on them. (They have this aversion to the sight of red ink!)

Would I teach the course this way again? Absolutely—the fact that they are unanimously interested in a writing in mathematics course makes me very happy indeed.

Acknowledgement. I would like to thank Dr. Richard Bullock, Director of Writing Programs at WSU, for his careful reading of this manuscript.

BIBLIOGRAPHY

1.   D. Byrkit, "Taxicab Geometry—a Non-Euclidean Geometry of Lattice Points," *Mathematics Teacher*, 1971: 418–422.

2.   M. Greenberg, *Euclidean and Non-Euclidean Geometries*, W. H. Freeman and Company, San Francisco, California, 1974.

3.   A. Hoffer, "Geometry is More Than Proof," *Mathematics Teacher*, 1981: 11–18.

4.   R. Millman and G. Parker, *Geometry: A Metric Approach with Models*, Springer-Verlag, New York, 1980.

5.   E. Moise, *Elementary Geometry from an Advanced Standpoint*, Addison-Wesley, Reading, Massachusetts, 1963.

6.   NCTM Yearbook, *Learning and Teaching Geometry, K–12*, Reston, VA, 1987.

7.   A. Young and T. Fulwiler, *Writing Across the Disciplines*, Boynton/Cook Publishers, Upper Montclair, New Jersey, 1986.

APPENDIX 1

1.   Are you?
     Undergraduate   7
     Graduate   5

2.   What grade did you receive in the course?

3.   Concerning the *suggested topics* list:

     a.   Was there enough of a choice?

          Yes   9   No   2

     b.   Were the topics interesting?

          Yes   9   No   1

     c.   Did the topic you picked turn out to be interesting?

          Yes   12   No   0

     d.   Was the need to use library books/journals valuable?

          Yes   12   No   0

     Comments:

4.   Concerning the *format* of the assignment:

     a.   Was requiring an initial draft useful?

          Yes   12   No   0

     b.   Was the length appropriate (about 5 pages)?

          Yes   12   Too long   0   Too short   0

     c.   Was the option to write multiple drafts with only the final result counting toward a grade valuable?

          Yes   12   No   0

     d.   Would more, but shorter, assignments be more valuable?

          Yes   3   No Different   4   No   3

     e.   Did you have access to and use word processing (e.g., a computer)?

Yes 7   No 5

5.   *Overall* effect of writing assignment:

   a.   Did you have more contact with the professor than you would have had without the writing assignment?

   Yes 6   No 4

   b.   Did you get a better idea of what geometry is about from the course because of the writing assignment?

   Yes 10   No 1

   c.   Did you learn much concerning writing about mathematics?

   a lot 5   a bit 6   none 1   negative effect 0

   d.   Would you be interested in an entire one quarter class in writing about mathematics?

   Yes 12   No 0

6.

   a.   The writing assignment was interesting.

   Yes 12   No 0

   b.   The writing assignment was a valuable part of the course.

   Yes 9   No Difference 0   No 1

   c.   The amount of time in hours that you spent on the writing assignment was 18 hours. (High of 45 hours, low of 6 hours.)

Overall Comments:

(Optional) Name:
Phone Number:

APPENDIX 2:   Paper Titles

1.   A Comparison of Taxicab Models—Burkit vs. Millman/Parker

2.   Basic Skills and Transformation Geometry

3.   Informal Geometry

4.   The World of the Taxicab

5.   Geometry, Spaces, and Rules

6.   Van Hiele Levels

7.   Introduction of Proofs in Geometric Logic

8.   Euclid

9.   A Framework for the Teaching of Geometry in High Schools

10.   Taxicab Geometry

11.   Studying Geometry from a Different Angle

12.   Non-Euclidean Geometry

13.   Geometry-Philosophical Implications

14.   A Summary of the Article and Applications of Ideas "Spadework Prior to Deduction in Geometry"

15.   Models in Abstract Art to Motivate Elementary Geometric Concepts

16.   Computer Graphics with the Aid of Geometry

17.   The Informal Approach to Teaching Geometry

18.   The Van Hiele Model

The Essay as a Cognitive Map: Results from a Course in the Foundations of Mathematics

James V. Rauff
Millikin University, Decatur, Illinois

Introduction

It is easy for undergraduate students of mathematics to develop simple stimulus-response conceptions of the content and process of mathematics. When faced with a problem, they will select an appropriate algorithm with which to solve the problem. Very rarely, however, do they construct some overall conception of how the various topics in a particular course form a coherent whole. This tendency towards compartmentalization is encouraged by the sectional subdivisions of most mathematics textbooks. Of course, as instructors we can supply a global view of the course content, but the students are rarely challenged to construct their own cognitive map of the course material.

In this essay I will relate the results of my attempt to force the students in my *Foundations of Mathematics* course to formulate their own cognitive map of the subject matter by writing a final comprehensive paper. The reader will, I am certain, see that the approach I have taken can easily be extended to other areas of mathematics.

The Course

The "Foundations of Mathematics" is an upper-division course for mathematics and computer science majors. In the semester from which the data in this paper are drawn there were two sophomore mathematics majors, one senior economics major, one senior industrial engineering major, one senior chemistry major, two senior computer science majors, one senior mathematics major, and six junior mathematics majors enrolled. Our textbook was William Hatcher's *The Logical Foundations of Mathematics* [2]. I also supplemented Hatcher with two essays on the philosophy of mathematics [3] and [1] and my own notes. The topics covered in the course included the rudiments of mathematical logic, Frege's formal system, Zermelo-Fraenkel set theory, Gödel's Incompleteness Theorem, and Turing computability. The students were required to read the textbook and the supplementary material; work technical exercises in proof theory, model theory, set theory, and recursive function theory; write critical essays on the readings; and present proofs in class. Students were rewarded for cogent critiques of each others work, which made for lively class meetings.

The Final Assignment

As a significant (67%) portion of their final exam, the students were required to write a comprehensive essay. The assignment given to them is reproduced here in its entirety:

"You have read about, heard about, and talked about many important concepts and results in the foundations of mathematics. Select 7 concepts (definitions), axioms, theorems, and/or theories which you consider to be the most significant or important.

Tell what each one is, what it says, why it is important to the foundations of mathematics, and how it relates to the others on your list. Do all of this in a coherent, typed (double-spaced) essay.

THERE IS NO SINGLE CORRECT ANSWER to this part of the assignment. I will assess your essay on its content and form. You need not pick the same seven items that I would. However, it is STRONGLY recommended that you draw from all the chapters of Hatcher that we covered."

The essay constituted 23% of their course grade. It was weighted heavily to insure their serious consideration. They had four weeks in which to complete the essay. During that period we discussed Turing computability and Hilbert's Tenth Problem (and its solution) in class.

The Results

The choices made by my students are given in Table 1. I have used the student's labels for the items because they reflect their understanding of the topics. Clearly, the items in the table are not mutually exclusive.

Table 1. *The students' view of the most significant topics in the foundations of mathematics.*

Topic	Number of Students
Gödel's Incompleteness Theorem	12
First Order Theories	10
Axiom of Choice (ZF)	10
Russell's Paradox	9
Zermelo-Fraenkel Set Theory	8
Frege's Formal System	7
Axiom of Replacement (ZF)	6
Axiom of Separation (ZF)	5
Peano Postulates	3
Axiom of Abstraction (Frege)	3
Turing Machines	2
Inconsistency	2
Tautology	2
Axiom of Extensionality	2
Intuitionism	2
First Order Logic	2
Criteria for Foundations	2
Recursive Functions	2
Gödel Numbers	1
Cardinal Numbers	1
Model (of a formal language)	1
Mathematical Induction	1
Relation	1
Orderings (on relations)	1
Axiom of Regularity (ZF)	1

Comments

I had two main goals in assigning this comprehensive essay. First, I wanted my students to think hard about how the separate topics in the course fit together as a whole. The best way to get one's thoughts clear, as we well know, is to put them down in writing. Second, I was interested in seeing what sort of cognitive map of the course each student had formed.

With respect to my first goal, I was pleased to find that most of the essays were well thought out and well written. (There were only two notable exceptions; the senior chemist who had already begun work at the Environmental Protection Agency, and a sophomore mathematics major who was in over his head and rejected all my suggestions that he remove himself from the course.)

I wanted my students to think hard about how the separate topics in the course fit together as a whole. The best way to get one's thoughts clear ... is to put them down in writing.

As I had hoped, the requirement that the topics be tied together in a whole package resulted in the students listing only those topics for which they had a clear understanding. The prominence of first order theories and the axiom of choice in the students' lists reflects the accessibility of these topics to undergraduates. As for the popularity of Gödel's Theorem (the proof of which is not trivial) and Russell's Paradox on the lists, my students were particularly taken by what they perceived as the negative power of these two results. As one junior mathematics major said about Russell's Paradox, "you can not just make up sets." The senior mathematics major, in discussing Gödel's Incompleteness Theorem writes (oversimplifying) "since we cannot prove everything (that is true), there has to be something that is missing."

When we look at the choices and their perceived interconnections as a representation of the students' cognitive map of the foundations of mathematics, several observations can be made. The most common technique used by the students was the narrative. Many found it easiest to present their list as a kind of Popperian sequence of conjectures and refutations. Russell's Paradox is seen as a refutation of Frege's system. Gödel's Theorem is seen as a refutation of the goals of formalism. The topics that do not fit nicely into the narrative become side plots or linking devices. The economics major criticized intuitionism for its disregard of the social process of mathematics and then moved to a discussion of the "social" theory building of Zermelo, Fraenkel, Church, and Turing.

Beyond the general observation that my students tend to see the topics in the foundations of mathematics in historical, processual terms (fueled no doubt by Hatcher's approach), we can see some reflection of their own particular interests in the choices they made. The two students that selected Turing machines for their lists were both computer science majors. They were both quite taken by Church's Thesis and the resulting connections between computer science and Gödel's results. Intuitionism was selected only by the economics major (who had studied in Europe, and had some sympathy for European ideas) and one of the junior mathematics majors who has a decidedly philosophical leaning. The concept of a model was listed only by the industrial engineering major. He emphasized the importance of the interrelationship between a model for a theory and the consistency of that theory. In his words, "a consistent theory must be a theory about something." This is an important result that I had expected more of my less practically oriented students to include.

Conclusion

What did this writing assignment do for me and for my students? Several of the students have remarked since turning in their papers that they had never had to try to "make sense" out of a whole mathematics course before. They had always concentrated on the algorithms, the definitions, the theorems, and the techniques of proof, but for the first time they had to come up with the "whole nine yards." I think this assignment helped them to increase their grasp of the specific topics, see the area as a whole, and recognize clearly those topics that they didn't really understand or see as purposeful.

As for me, reading these papers has given me my clearest view of how my students organize the topics in the foundations of mathematics. That only a handful presented the material in a fashion which coincided with my view can be seen both positively and negatively. Clearly, the students actually formulated their own cognitive maps. However, does that mean that my overall plan for the course was obscure or not clearly expressed? For me, the positive aspect of fourteen students constructing and defending their own understanding of the foundations of mathematics far outweighs the negative.

They had never had to try to "make sense" out of a whole mathematics course before.

The comprehensive organizational essay is a good way to get the students to crystallize their cognitive maps of a course and at the same time give feedback to the instructor on what was deemed important, what was truly understood, and what was neither.

REFERENCES

1. N. Goodman, "Mathematics as an Objective Science," *American Mathematical Monthly*, 86 (1979): 540–551.

2. W. Hatcher, *The Logical Foundations of Mathematics*, Pergamon Press, Oxford, 1982.

3. C. Hempel, "On the Nature of Mathematical Truth," in H. Feigl and M. Brodbeck (eds.), *Readings in the Philosophy of Science*, Appleton-Century Crofts, New York, 1953: 148–162.